15. Band, 1. Heft

Fortschritte der chemischen Forschung
Topics in Current Chemistry

Springer-Verlag   6900 Heidelberg 1 · Postfach 1780
Telefon (06221) 49101 · Telex 04-61723
1000 Berlin 33 · Heidelberger Platz 3
Telefon (0311) 822001 · Telex 01-83319

Springer-Verlag   New York, NY 10010 · 175, Fifth Avenue
New York Inc.   Telefon 673-2660

Theory of Orientation and Stereoselection

Prof. K. Fukui

Kyoto University, Department of Hydrocarbon Chemistry, Kyoto, Japan

Contents

1. Molecular Orbitals

Many chemical problems can be discussed by way of a knowledge of the electronic state of molecules. The electronic state of a molecular system becomes known if we solve the electronic *Schrödinger equation*, which can be separated from the time-independent, nonrelativistic Schrödinger equation for the whole molecule by the use of the Born-Oppenheimer approximation [1]. In this approximation, the electrons are considered to move in the field of momentarily fixed nuclei. The nuclear configuration provides the parameters in the Schrödinger equation.

The nonrelativistic, electronic *Schrödinger Hamiltonian operator*, designated as H, is represented by

$$H = \sum_{\substack{i=1}}^{N} \left(-\frac{h^2}{8\pi^2 m} \Delta_i - \sum_a \frac{Z_a e^2}{r_{ia}} \right) + \sum_{\substack{i,j=1 \\ (i<j)}}^{N} \frac{e^2}{r_{ij}} + \sum_{\substack{a,b \\ (a<b)}} \frac{Z_a Z_b e^2}{r_{ab}} \qquad (1.1)$$

in which

N is the number of electrons,

Δ_i is the Laplacian operator for electron i,

$Z_a e$ is the positive charge of nucleus, a, and

r_{ij}, r_{ia}, and r_{ab} are the distances between electrons i and j, nucleus a and electron i, and nuclei a and b,

respectively: e and m are the charge and the mass of an electron: h is the Planck constant.

The eigenstate of the operator H may be described in terms of $4N$ electron coordinates,

$$x_i, y_i, z_i, \quad \text{and} \quad \xi_i \ (i = 1, 2, \text{-----} N),$$

where the first three are the Cartesian coordinates and the last one is the spin coordinate. The *wave function*, Ψ, of an eigenstate of H is therefore represented by $\Psi(12 \text{-----} N)$ in which $i \, (i = 1, 2, \text{-----} N)$ stands for the set of coordinates (x_i, y_i, z_i, ξ_i).

From the well-known statistical requirement for an assembly of Fermi particles, $\Psi(12 \text{-----} N)$ is subject to a limitation in its form of antisymmetric character with respect to electron exchange. In addition to this, we have to note that an eigenstate of H can be specified also by the *eigenvalues* of $\mathbf{S}^2$ and S_z, where $\mathbf{S}$ is the *total electronic spin angular momentum vector*. In this way, we are able to obtain information about the general form which should be satisfied by the simultaneous eigenfunction of H, $\mathbf{S}^2$, and S_z. Let such a function be denoted by Ψ_{SM_S} in which S and M_S specify the eigenvalues of $\mathbf{S}^2$ and S_z, respectively. In this way, the form which must be taken by an antisymmetric spin-eigenstate N-electron wave function can be derived.

For instance, as is well known, the general form of wave functions with $N = 2$, $S = 0$, $M_S = 0$ is

$$\{\psi(12) + \psi(21)\}\{\alpha(1)\beta(2) - \beta(1)\alpha(2)\} \tag{1.2}$$

where $\psi(12)$ is an arbitrary two-electron spatial function, and α and β are the usual spin functions. If an "exact" eigenfunction of H for a two-electron system were obtained, it would naturally be of this form.

Such a "general" form of wave function is easily written explicitly for each set of values of N, S, and M_S. Any appropriate form of approximate wave functions, like determinantal functions composed of one-electron functions ("molecular spin orbitals"), the "bond eigenfunctions" used in the valence bond approach, and so on, is shown to fulfil this requirement.

Some of these approximate forms of wave function possess a character of particular theoretical interest. One such is the *"uni-configurational"* *wave function*. This implies an appropriate linear combination of antisymmetrized products of molecular spin orbitals in which all antisymmetrized products belong to the same "electron configuration". The electron configuration of an antisymmetrized product is defined as the set of N spatial parts appearing in the product of spin orbitals. For instance, a uni-configurational wave function with $N = 2$, $S = 0$, $M_S = 0$ is expressed as

$$(i\bar{j}) - (\bar{i}j) \tag{1.3}$$

where

$$(i\bar{j}) \equiv \begin{vmatrix} i(1)\,\alpha(1)\,j(1)\,\beta(1) \\ i(2)\,\alpha(2)\,j(2)\,\beta(2) \end{vmatrix} \quad \text{etc.,}$$

and the set $[ij]$ stands for the electron configuration. The spatial part of a spin orbital is often called simply an *"orbital"*. The orbital which appears only once in an electron configuration is said to be "singly occupied", and that appearing twice "doubly occupied".

The general form of such uni-configurational wave functions can be obtained for any set of N, S, and M_S. It is easy to see that such a form of wave functions duly satisfies the general requirement mentioned above, as in Eq. (1.2).

Some uni-configurational wave functions consist of only one determinant. This is called a *single-determinant wave function*. A single-determinant can be a spin-eigenstate wave function only if the eigenfunctions possess the values of

$$S = |M_S| = \tfrac{1}{2}(N - 2\nu)$$

where ν is the number of doubly occupied orbitals in the determinant. Thus

[*case A*] open-shell wave functions with maximum multiplicity $(\nu = 0, S = |M_S| = N/2)$,

[*case B*] closed-shell wave functions $(\nu = N/2, S = |M_S| = 0)$, and

[*case C*] wave functions with a closed-shell structure of ν doubly occupied orbitals with additional open-shell structure of $S = |M_S| = \tfrac{1}{2}(N - 2\nu)$ belong to this category. Any other uni-configurational wave functions consist of more than one determinant.

We can discuss the "best" uni-configurational wave function by the usual variational method of the *Hartree-Fock* type. This means making a search for the function Ψ which minimizes the quantity

$$\int \Psi^* H \Psi \, d\tau / \int \Psi^* \Psi \, d\tau . \tag{1.4}$$

If an excited state is concerned, this is done under the restriction that the function should be orthogonal to all of the lower-energy states. We may specify these as the *"uni-configurational Hartree-Fock wave functions"*. The "best" orbitals constructing the determinants in these wave functions are in general not orthogonal to each other.

In [*case A*] and [*case B*] mentioned above, the "best" wave function thus obtained is of particular practical importance. The set of N orbitals appearing in these functions is in general definitely determined, except for an arbitrary numerical factor of which the absolute value is unity, as being mutually orthogonal and having a definite "orbital energy" [cf.

5

Eq. (3.15)]. The concept of "electron occupation" of orbitals is thus unequivocal in these cases. The best orbitals in these cases are called "Hartree-Fock orbitals"[2,3].

The wave function of [*case A*] is in general written in the form

$$\frac{1}{\sqrt{N!}} \begin{vmatrix} \phi_1(1) & \phi_2(1) & \text{------} & \phi_N(1) \\ \phi_1(2) & \phi_2(2) & \text{------} & \phi_N(2) \\ \text{---------------------} \\ \phi_1(N) & \phi_2(N) & \text{----} & \phi_N(N) \end{vmatrix} \sigma^{(S)}(1, 2, \text{-----} N) \qquad (1.5)$$

where $\phi_i(k)$ is the ith orbital occupied by the *kth* electron and $\sigma^{(S)}(1, 2, \text{-----} N)$ is the totally symmetric N-electron spin function.

The wave function of [*case B*] with $N=2$ can be written as

$$\phi_1(1) \phi_1(2) \frac{1}{\sqrt{2}} \{\alpha(1) \beta(2) - \beta(1) \alpha(2)\} \qquad (1.6)$$

The closed-shell wave functions with $N > 2$ can no longer be separated into spatial and spin parts, but are expressed in the following form:

$$\frac{1}{\sqrt{(2\nu)!}} \begin{vmatrix} \phi_1(1)\alpha(1) \phi_1(1)\beta(1) \phi_2(1)\alpha(1) \phi_2(1)\beta(1) \text{---} \phi_\nu(1)\alpha(1) \phi_\nu(1)\beta(1) \\ \phi_1(2)\alpha(2) \text{------------------------------} \phi_\nu(2)\beta(2) \\ \text{---} \\ \phi_1(N)\alpha(N) \text{------------------------------------} \phi_\nu(N)\beta(N) \end{vmatrix} \qquad (1.7)$$

Such a determinantal form of wave function is often called the Slater determinant.

Thus, we have the N-electron wave function with separated spatial and spin parts only in the cases of two-electron singlet states and N-electron $(N+1)$-plet states. The Hartree-Fock orbitals are defined as those functions ϕ_i which make the wave functions (1.5), (1.6), and (1.7) best. The usual variation technique leads to the N(case A) or ν(case B) simultaneous differential equations which have to be satisfied by $\phi_i (i = 1, 2, \text{---} N$ in case A, and $i = 1, 2, \text{---} \nu$ in case B). These equations are called the Hartree-Fock equations. The Hartree-Fock orbitals are obtained by solving these differential equations simultaneously.

Besides the occupied orbitals, these equations possess solutions corresponding to actually unoccupied, virtual orbitals. Some of them happen to possess negative energies (corresponding to "bound one-electron states"), whereas the others have nonnegative energies. The Hartree-Fock unoccupied orbital, rather than its realistic physical meaning, is important in the sense that it is used in *constructing excited-state wave functions* and plays a significant role in the theory of chemical interactions (Chap. 3). It is to be remarked that the mathematical means suitable for describing the unoccupied orbitals are not always the same as those representing the occupied orbitals with tolerable approximation.

The Hartree-Fock equations for the *hydrogen molecule* have been solved by Kolos and Roothaan[4], by obtaining the binding energy value of 3.63 eV for the ground state, which is ca. 1.1 eV smaller than the exact theoretical value [4,5]. This difference corresponds to the correlation error. The Hartree-Fock orbital energies of other *homonuclear diatomic molecules*, C_2, N_2, O_2 and F_2, have been obtained by Buenker *et al.* [6]. A review has been given by Wahl *et al.* [7] with illustrative orbital maps for the F_2, NaF, and N_2 molecules. Also calculations have been made with respect to simple *hydrocarbons* such as CH_4, C_2H_6, C_2H_4, and C_2H_2 [6,8,9].

The Hartree-Fock orbitals are expanded in an infinite series of known basis functions. For instance, in diatomic molecules, certain two-center functions of elliptic coordinates are employed. In practice, a limited number of appropriate atomic orbitals (AO) is adopted as the basis. Such an approach has been developed by Roothaan [10]. In this case the Hartree-Fock differential equations are replaced by a *set of nonlinear simultaneous equations* in which the limited number of AO coefficients in the linear combinations are unknown variables. The orbital energies and the AO coefficients are obtained by solving the Fock-Roothaan secular equations by an iterative method. This is the procedure of the Roothaan LCAO (linear-combination-of-atomic-orbitals) SCF (self-consistent-field) method.

The basis AO adopted may be Slater-type AO [11], Gaussian AO [12], and Hartree-Fock AO [13], Löwdin's orthogonalized AO [14], and so on. In many cases the Slater AO's for the valence-shell electrons are taken. Clementi has extended the basis beyond the valence shells [15]. Frequently, the exponents of Slater AO's are optimized. Clementi has also adopted two different variable exponents for "one" Slater AO [15].

Even an exact Hartree-Fock calculation cannot be exempt from the correlation error. A practical method of evaluation has been proposed by Hollister and Sinanoğlu [16]. An *LCAO SCF method* has been applied to the calculation of the heat of various simple reactions by Snyder and Basch [17]. They have evaluated the correlation error by the method of Hollister and Sinanoğlu [16].

In the cases other than [*case A*] and [*case B*], so called *"open-shell"* *SCF methods* are employed. The orbital concept becomes not quite certain. The methods are divided into classes which are "restricted" [18] and "unrestricted" [19] Hartree-Fock procedures. In the latter case the wave function obtained is no longer a spin eigenfunction.

The Hartree-Fock method is modified by mixing some important valence electron configurations with the ground-state one [20]. This is called the *OVC (optimized valence configurations) method*.

Such a wave function is represented by a linear combination of wave functions for more than one electron configuration, and is called a *"multi-configurational" wave function*. The consideration of more than one configuration can reduce the correlation error. Such an approach is referred to as the *method of "configuration interaction* (CI)*"*.

Some useful, conventional SCF methods have been proposed by Pople [21] and by Kon [22] using the semiempirical calculation of Pariser and Parr [23] with regard to the π electrons of planar conjugated molecules.

Yonezawa *et al.* [24] have developed an SCF method taking into account all valence electrons with all overlap integrals included. They have made calculations with respect to several simple molecules, such as

CH_4, C_2H_6, C_2H_4, C_2H_2, CO, CO_2, H_2O, H_2CO, CH_3OH, HCN, and NH_3 [24];

larger molecules like butadiene, acrolein, and glyoxal [25]; several alkyl radicals of $C_1 \sim C_4$ [26]; and aza-heterocycles [27]. This method gives reasonable theoretical values for transition energies, ionization potentials, dipole moments, and chemical reactivities of these molecules.

A method which is similar to the Pariser-Parr-Pople method for the π electron system and is applicable to common, saturated molecules has been proposed by Pople [28]. This method is called the *CNDO (complete neglect of differential overlap) SCF calculation*. Katagiri and Sandorfy [29] and Imamura *et al.* [30] have used hybridized orbitals as basis of the Pariser-Parr-Pople type semiempirical SCF calculation.

Other approximate, more empirical methods are the extended *Hückel* [31] *and hybrid-based Hückel* [32,33] *approaches*. In these methods the electron repulsion is not taken into account explicitly. These are extensions of the early Hückel molecular orbitals [34] which have successfully been used in the π electron system of planar molecules. On account of the simplest feature of calculation, the Hückel method has made possible the first quantum mechanical interpretation of the classical electronic theory of organic chemistry and has given a reasonable explanation for the chemical reactivity of sizable conjugated molecules.

2. Chemical Reactivity Theory

From 1933 [35], several theoretical approaches to the problem of the chemical reactivity of planar conjugated molecules began to appear, mainly by the Hückel molecular orbital theory. These were roughly divided into two groups [36]. The one was called the *"static approach"* [35,37—40], and the other, the *"localization approach"* [41,42]. In 1952, another method which was referred to as the *"frontier-electron method"* was proposed [43] and was conventionally grouped [44] together with other related methods [45,46] as the *"delocalization approach"*.

The first paper of the frontier-electron theory pointed out that the *electrophilic aromatic substitution* in aromatic hydrocarbons should take place at the position of the greatest density of electrons in the *highest occupied* (HO) molecular orbital (MO). The second paper disclosed that the nucleophilic replacement should occur at the carbon atom where the *lowest unoccupied* (LU) MO exhibited the maximum density of extension. These particular MO's were called "frontier MO's". In homolytic replacements, both HO and LU were shown to serve as the frontier MO's. In these papers the "partial" density of $2 p\pi$ electron, in the HO (or LU) MO, at a certain carbon atom was simply interpreted by the square of the atomic orbital (AO) coefficient in these particular MO's which were represented by a linear combination (LC) of $2 p\pi$ AO's in the frame of the Hückel approximation. These partial densities were named "frontier-electron densities".

The explanation of these findings was at that time never self-evident. In contrast to the other reactivity theories, which then existed and had already been well-established theoretically, the infant frontier-electron theory was short of solid physical ground, having suggested a possibility of the involvement of a new principle relating to the nature of chemical reactions.

In the same year as that of the proposal of the frontier-electron theory, the *theory of charge-transfer force* was developed by Mulliken with regard to the molecular complex formation between an electron donor and an acceptor [47]. In this connection he proposed the "overlap and orientation" principle [48] in which only the overlap interaction between the HO MO of the donor and the LU MO of the acceptor is considered.

The behaviour of the frontier electrons was also attributed to a certain type of electron delocalization between the reactant and the reagent [49]. A concept of *pseudo-π-orbital* was introduced by setting up a simplified model, and the electron delocalization between the π-electron system of aromatic nuclei and the pseudo-orbital was considered to be essential to aromatic substitutions. The pseudo-orbital was assumed to be built up out of the hydrogen atom AO attached to the carbon atom at the reaction center and the AO of the reagent species, and to be occupied by zero, one, and two electrons in electrophilic, radical, and nucleophilic reactions. A theoretical quantity called "superdelocalizability" was derived from this model. This quantity will be discussed in detail later in Chap. 6.

a) Reaction with an
   electrophilic reagent

b) Reaction with a
   radical reagent

c) Reaction with a
   nucleophilic reagent

π-System Pseudo-orbital   π-System Pseudo-orbital   π-System Pseudo-orbital

The frontier-electron density was used for discussing the reactivity within a molecule, while the superdelocalizability was employed in comparing the reactivity of different molecules [44]. Afterwards, the applicability of the frontier-electron theory was extended to saturated compounds [50]. The new theoretical quantity "delocalizability" was introduced for discussing the reactivity of saturated molecules [50]. These indices satisfactorily reflected experimental results of various chemical reactions. In addition to this, the conspicuous behavior of HO and LU in determining the steric course of organic reactions was disclosed [44,51].

All of these facts make one believe that the distinction of particular MO's, the frontier orbitals, from the others has a good reason which arises from the general principle governing the nature of chemical reactions. It is useful in this connection to analyze first the interaction energy of two reacting species in general [52]. The energy is divided into several terms so that one can understand what kind of interaction energy is really important in chemical reactions.

3. Interaction of Two Reacting Species

Two isolated reactant molecules in the closed-shell ground state are designated as A and B, whose electronic energies are W_{A0} and W_{B0}, respectively. Here the term closed-shell implies the structure of a molecule with doubly occupied MO's only. The lowest total energy of the two mutually interacting systems is denoted by W. Then, the interaction energy is defined by

$$\Delta W = W - (W_{A0} + W_{B0}) \tag{3.1}$$

All the energy values are calculated by the Born-Oppenheimer approximation with respect to a fixed nuclear configuration. The most stable configurations of interacting systems are obviously different from the respective isolated systems. However, the nuclear configuration change is tentatively left untouched in order to disclose the constitution of interaction energy at the beginning of the theory. Namely, W is the energy of a system composed of A and B approaching each other without deformation, satisfying the Schrödinger equation for the combined system

$$H \Psi = W \Psi \tag{3.2}$$

in which the Hamiltonian operator H is represented by

$$H = \sum_{\lambda} H(\lambda) + \sum_{\lambda < \lambda'} \frac{e^2}{r_{\lambda\lambda'}} + \sum_{\gamma < \gamma'} \frac{Z_\gamma Z_{\gamma'} e^2}{R_{\gamma\gamma'}} \tag{3.3}$$

$$H(\lambda) = - \frac{h^2}{8\pi^2 m} \Delta(\lambda) + V(\lambda) \tag{3.4}$$

$$V(\lambda) = V_A(\lambda) + V_B(\lambda) \tag{3.5}$$

$$V_A(\lambda) = - \sum_a \frac{Z_a e^2}{r_{\lambda a}}, \qquad V_B(\lambda) = - \sum_\beta \frac{Z_\beta e^2}{r_{\lambda\beta}} \tag{3.6}$$

$H(\lambda)$ is the one-electron Hamiltonian operator of the electron λ

Z_α, Z_β, and Z_γ are the positive charge numbers of the nuclei α, β, and γ, belonging to molecule A, molecule B, and the combined system, AB

$r_{\lambda\lambda'}$ is the distance between the two electrons λ and λ':

$r_{\lambda\alpha}$ is the distance of the electron λ from the fixed nucleus α

$R_{\alpha\alpha'}$ is the distance between the fixed nuclei α and α'

$\Delta(\lambda)$ is the Laplacian operator for the electron λ

$V_A(\lambda)$ and $V_B(\lambda)$ are the potential energies of the electron λ due to the nuclei belonging to molecules A and B, respectively.

To compose the wave function Ψ for the combined system $A\text{-----}B$, an attempt is made to employ the MO's of the isolated reactant molecules A and B. The unperturbed normalized wave functions of A and B are represented in terms of the Slater determinants composed of ortho-normal (mutually orthogonal (cf. Chap. 1) and normalized) spin orbitals. The spin orbitals are assumed to have the spatial parts which are made SCF MO with respect to the *ground state* of each isolated molecule, A or B, in the Hartree-Fock sense (Chap. 1). To make an approximate excited-state wave function of an isolated system, the Hartree-Fock unoccupied MO's mentioned in Chap. 1 which are associated with the Hartree-Fock equation for the ground state are employed in constructing the Slater determinant. In this way, all of the MO's which are used in the wave function for the combined system $A\text{-----}B$ are defined definitely with regard to a given nuclear configuration in each isolated system.

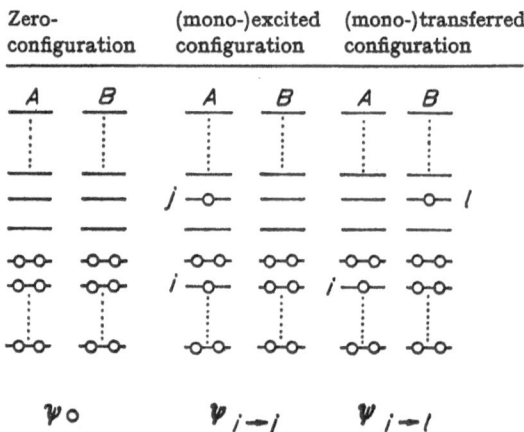

Fig. 3.1. The electron configuration of the combined system $A\text{-----}B$

The wave function Ψ for the combined system $A-----B$ is represented by a multi-configurational one which is a linear combination of the spin-eigenstate determinantal functions composed of the above-defined spin orbitals, which are antisymmetrized with respect to all electrons of the whole system. These determinantal functions correspond to the *electron configurations* illustrated below.

The zero-configuration corresponds to the combined system in which A and B interact in their ground state. In an excited configuration either A or B (or both) is in an excited state. The transferred configuration is one in which one (or more) electron is transferred from an MO of one system to an MO of the other. The MO's occupied and unoccupied in the ground state are discerned by the following notation:

The wave function is thus represented by

$$\Psi = C_0 \Psi_0 + \left(\underset{p}{\overset{\text{monoex.}}{\sum}} + \underset{p}{\overset{\text{monotr.}}{\sum}} + \underset{p}{\overset{\text{diex.}}{\sum}} + \underset{p}{\overset{\substack{\text{monoex.-}\\\text{monotr.}}}{\sum}} + \underset{p}{\overset{\text{ditr.}}{\sum}} + --- \right) C_p \Psi_p \qquad (3.7)$$

where suffix 0 implies the zero-configuration, *ex.* and *tr.* signify *"excited"* and *"transferred"*, respectively, Ψ_p represents one of the wave functions $\Psi_{i \to j}$, $\Psi_{i \to l}$, etc. corresponding to the above-depicted electron configurations, or highly excited or transferred ones, and C_0 and C_p are coefficients which are to be determined so as to minimize the total energy of the combined system $A-----B$.

An approach such as this belongs to the method of configuration interaction (CI) mentioned in Chap. 1. It is sufficient to cite a simple example to illustrate the usefulness of such CI treatments. It is well known that the Weinbaum wave function [53] for the hydrogen molecule gives a better result than the Hartree-Fock calculation, notwithstanding the simplest form as follows:

$$\Phi(1,2) = a\left\{\chi_A(1)\,\chi_B(2) + \chi_B(1)\,\chi_A(2)\right\} + b\left\{\chi_A(1)\,\chi_A(2) + \chi_B(1)\,\chi_B(2)\right\}$$

where χ_A and χ_B are the 1s AO of the hydrogen atoms A and B with the effective nuclear charge larger than unity ($=1.193$). This implies that the AO is "shrunken". Mulliken [54] has shown that the 1s function with orbital exponent 1.2 can be expanded in terms of ns functions with orbital exponent unity:

$$(1s)(\zeta = 1.2) = 0.9875(1s)(\zeta = 1) - 0.0925(2s)(\zeta = 1)$$
$$+ 0.0433(3s)(\zeta = 1) -----$$

On substituting such an expansion into the wave function formula, it becomes evident that this function consists of the following terms:

which corresponds obviously to Eq. (3.7). The success of Weinbaum's treatment may be attributed to the CI nature of that treatment.

In view of the impracticability of the Hartree-Fock calculation for common molecules, the LCAO MO spatial functions may be used in place of Hartree-Fock ones. The MO's a and b are given by

$$a(1) = \sum_t c_t\, t(1) \qquad \text{for molecule } A$$

$$b(1) = \sum_u c_u\, u(1) \qquad \text{for molecule } B \tag{3.8}$$

where 1 implies the coordinates of the electron 1, and t and u are the AO's belonging to the nuclei of A and B, respectively. The coefficients c_t and c_u are chosen so that $a(1)$ and $b(1)$ become Roothaan-type SCF MO's [10] for the ground state of each isolated system. The AO's $t(1)$ and $u(1)$ may usually be taken to be real Slater-type AO's, for example.

The total energy W in Eq. (3.2) is obtained by solving the usual secular equation [52] as

$$W = H_{0,0} - \left(\overset{\text{monoex.}}{\underset{p}{\sum}} + \overset{\text{monotr.}}{\underset{p}{\sum}} + \overset{\text{diex.}}{\underset{p}{\sum}} + \overset{\substack{\text{monoex.-}\\\text{monotr.}}}{\underset{p}{\sum}} + \overset{\text{ditr.}}{\underset{p}{\sum}} + \text{---} \right)$$

(3.9)

$$\frac{|H_{0,p} - S_{0,p} H_{0,0}|^2}{H_{p,p} - H_{0,0}} + \text{---}$$

where

$$H_{p,q} = \int \Psi_p^* H \Psi_q \, d\tau \quad \text{and} \quad S_{p,q} = \int \Psi_p^* \Psi_q \, d\tau$$

and the wave function Ψ in Eq. (3.7) is simultaneously determined.

Consider the case where the interaction between the molecules A and B is not yet very strong. The magnitude of $H_{0,p}$ is almost linear with $S_{0,p}$, so that the second-order term in Eq. (3.9) is proportional to the square of $S_{0,p}$. The order of magnitude of $S_{0,p}$ is equal to the νth power of an overlap integral s_{ab} of an MO a of the molecule A and an MO b of the molecule B, where γ is the minimum number of electron transfers between A and B required to shift the electron configuration from 0 to p. Therefore, the terms from monotransferred configurations in Eq. (3.9) have magnitudes of the order of s_{ab}^2, while the monoex. and the ditr. terms are of s_{ab}^4, and the monoex.-monotr. term s_{ab}^6, the diex. term s_{ab}^8, and so on. If the interaction is weak and s_{ab} is small, the mono-transferred terms are important in comparison with the others.

There are some additional reasons which make the contribution of *monotransferred terms* uniquely important. As assumed before, the MO's used are the Hartree-Fock or other SCF ones so that the values of $H_{0,p}$ of monoex. terms are small, since the Brillouin theorem [55] requires that the matrix element between the ground state and a monoexcited state in the Hartree-Fock approach should vanish in an isolated molecule. In addition to this, the denominator of the second-order term

$(H_{p,q} - H_{0,0})$ in Eq. (3.9) can usually not be small in excited configuration terms, whereas in transferred configuration terms it can be. Even a first-order term of the form

$$- | H_{0,p} - S_{0,p} H_{0,0} | \tag{3.10}$$

appears in place of the second-order term

$$- \frac{| H_{0,p} - S_{0,p} H_{0,0} |^2}{H_{p,p} - H_{0,0}}$$

when $H_{p,p}$ is approximately equal to $H_{0,0}$, that is, in a "degenerate" case. From these considerations, the following approximate formula is obtained:

$$W \simeq H_{0,0} - \sum_{p}^{\text{monoex.}} \frac{| H_{0,p} - S_{0,p} H_{0,0} |^2}{H_{p,p} - H_{0,0}} \tag{3.11}$$

The interaction energy, ΔW, in Eq. (3.1) is in this way converted into the form

$$\Delta W \simeq \varepsilon_Q + \varepsilon_K - D \tag{3.12} \text{ [56]}$$

where ε_Q is the Coulomb interaction term represented by

$$\varepsilon_Q \simeq \sum_{\alpha} \sum_{\beta} e^2 \frac{(Z_\alpha - N_\alpha)(Z_\beta - N_\beta)}{R_{\alpha\beta}} \tag{3.13}$$

by the use of Mulliken's approximation [57], in which N_α is the population of electrons, so that $e(Z_\alpha - N_\alpha)$ is the net plus charge, of the atom α, ε_K is the *exchange interaction term*, and D is the stabilization energy due to the charge-transfer interaction, which is written in the following form

$$D = \sum_{i}^{\text{occ}} \sum_{l}^{\text{uno}} \frac{|H_{0,i \to l} - S_{0,i \to l} H_{0,0}|^2}{H_{i \to l, i \to l} - H_{0,0}} + \sum_{k}^{\text{occ}} \sum_{j}^{\text{uno}} \frac{|H_{0,k \to j} - S_{0,k \to j} H_{0,0}|^2}{H_{k \to j, k \to j} - H_{0,0}} \tag{3.14}$$

where $\overset{\text{occ}}{\sum}$ and $\overset{\text{uno}}{\sum}$ imply the summation covering the occupied and unoccupied MO's, respectively.

The form of Eq. (3.13) indicates that this term is the sum of Coulomb potentials arising from the net charge of each atom of molecule A and that of each atom of molecule B. Therefore, ε_Q is significant in the interaction of polar molecules, causing a long-range force.

The *exchange interaction term*, ε_K, is important in the short range, being as usual repulsive in the interaction of closed-shell molecules, although it behaves as attractive in the singlet interaction of two odd-electron systems. Suppose that the overlapping of MO's of A and B takes place appreciably only between one AO, say r, of A and one AO, say r', of B. Such a mode of interaction may be called single-site overlapping, and is nearly realized in the aromatic substitution by a reagent with essentially one AO. In such cases the exchange interaction terms vary with the square of the overlap integral $s_{rr'}$, so that they are less important than the Coulomb term, at least at the initial stage of interaction of two closed-shell molecules.

The term D of Eq. (3.14) is called the *delocalization stabilization*, which is usually positive. This term comes from the electron delocalization between the molecules A and B. The physical meaning of the denominator of each term in the right side of Eq. (3.14) can be discussed in relation to the Koopmans theorem [58]

$$\varepsilon_i = -I_i \tag{3.15}$$

in which ε_i is the energy of the ith MO and I_i is the ionization potential with respect to the electron in the ith MO. From the result of calculation [52] it follows that

$$H_{i \to l, i \to l} - H_{0,0} = I_{Ai}^{(B)} - E_{Bl}^{(A-i)} \tag{3.16}$$
$$= I_{Ai}^{(B+l)} - E_{Bl}^{(A)}$$

where I_{Ai} is the ionization potential of A with respect to the ith MO and E_{Bl} is the electron affinity of B with respect to the lth MO, and $I_{Ai}^{(B)}$ signifies the I_{Ai} value in the case of the approach of molecule B, $E_{Bl}^{(A)}$ is the value of E_{Bl} with the approach of molecule A, $I_{Ai}^{(B+l)}$ is the I_{Ai} in the approach of molecule B with an additional electron in the lth MO which is unoccupied in the ground state, and $E_{Bl}^{(A-i)}$ is the value of E_{Bl} in the case of the approach of molecule A in which one electron in the

ith MO is subtracted. The relation of Eq. (3.16) is schematically represented by the following figure:

$$H_{i-1, i-1} - H_{0,0}$$

$$= I_{Ai}^{(B)} - E_{Bi}^{(A-i)}$$

$$= I_{Ai}^{(B+1)} - E_{Bi}^{(A)}$$

The integrals appearing in the numerator of each term in the right side of Eq. (3.14) can be rewritten as [52)]

$$H_{0, i \to l} - S_{0, i \to l} H_{0,0} = 2 \sum_{r} c_r^{(l)} c_{r'}^{(l)} \gamma_{rr'}^{(l)}, \qquad (3.17)$$

in which the multiple-site interaction between molecules A and B is assumed to take place through a paired overlapping of the rth AO of A and the r'th AO of B,

where

$$\gamma_{rr'}^{(l)} \cong - \int r(1) \left(\sum_{\beta} \frac{Z_\beta - N_\beta}{r_{1\beta}} \right) r'(1) \, dv(1) + s_{rr'} \sideset{}{'}\sum_{a} \sideset{}{'}\sum_{\beta} \frac{n_a^{(ll)} (Z_\beta - N_\beta)}{R_{a\beta}} \qquad (3.18)$$

and

$$n_a^{(ll)} = \sideset{}{^{(a)}}\sum_{t} \sum_{t'} c_t^{(l)} c_{t'}^{(l)} s_{tt'} \qquad (3.19)$$

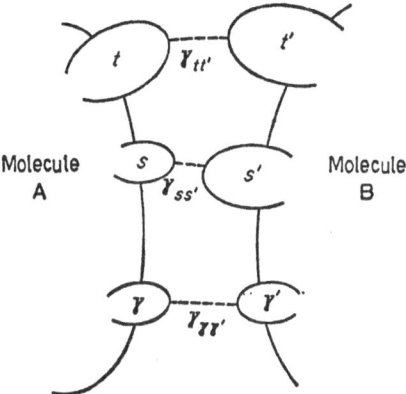

Fig. 3.2. The mode of multiple-site overlapping interaction

in which $c_t^{(i)}$ is the coefficient of the tth AO in the ith MO as in Eq. (3.8.) Eq. (3.18) indicates that the charge-transfer interaction is governed by the local net charge near the reaction center of the acceptor molecule.

In this way, the expression for the delocalization stabilization, D, is obtained as

$$D \sim 2 \left\{ \sum_i^{occ} \sum_l^{uno} \frac{(\sum_r c_r^{(i)} c_{r'}^{(l)} \gamma_{rr'}^{(i)})^2}{I_{Ai}^{(B)} - E_{Bl}^{(A-i)}} + \sum_k^{occ} \sum_j^{uno} \frac{(\sum_r c_r^{(j)} c_{r'}^{(k)} \gamma_{rr'}^{(k)})^2}{I_{Bk}^{(A)} - E_{Aj}^{(B-k)}} \right\} \qquad (3.20)$$

This quantity represents the energy of the multiple-site charge-transfer interaction which will later play an important role in the theory of stereoselection. It is to be remarked that, although any MO may involve an arbitrary constant of which the absolute value is unity, the value of the numerator in each term of the right side of this equation is always definite.

One of the most important special cases is that of the single-site interaction between the rth AO of the reactant, A, and a reagent, B, which possesses only one AO designated as r'. In this case D is written as

$$D \sim \begin{cases} 2 \sum_i^{occ} \frac{c_r^{(i)2}}{\varepsilon_B - \varepsilon_{Ai}} \gamma_r^2 & \text{(the reagent orbital is unoccupied)} \quad (3.21\,a) \\[2em] 2 \sum_j^{uno} \frac{c_r^{(j)2}}{\varepsilon_{Aj} - \varepsilon_B} \gamma_r'^2 & \text{(the reagent orbital is occupied)} \quad (3.21\,b) \end{cases}$$

where

$$\gamma_r = c_r^{(l)} \gamma_{rr'}^{(i)}, \quad \gamma_r' = c_r^{(k)} \gamma_{rr'}^{(k)}, \quad \varepsilon_{Ai} = -I_{Ai}^{(B)}, \quad \varepsilon_{Aj} = -E_{Aj}^{(B-k)},$$

and

$$\varepsilon_B = \begin{cases} -E_B^{(A-i)} & \text{(the reagent orbital is unoccupied)} & (3.22\,\text{a}) \\[2mm] -I_B^{(A)} & \text{(the reagent orbital is occupied)} & (3.22\,\text{b}) \end{cases}$$

The right sides of Eq. (3.21) can be employed as a measure of the chemical reactivity of both saturated and unsaturated compounds, which will be discussed in detail later.

The case of interaction between an *even-electron molecule A* and an *odd-electron molecule B* can be discussed in a similar manner. Eq. (3.20) is modified to be

$$D \sim 2 \left\{ \sum_i^{\text{occ}} \sum_l^{\text{uno}} \frac{\left(\sum_r c_r^{(i)} \cdot c_{r'}^{(l)} \gamma_{rr'}^{(i)} \right)^2}{I_{Ai}^{(B)} - E_{Bl}^{(A-i)}} + \sum_k^{\text{occ}} \sum_j^{\text{uno}} \frac{\left(\sum_r c_r^{(j)} c_{r'}^{(k)} \gamma_{rr'}^{(k)} \right)^2}{I_{Bk}^{(A)} - E_{Aj}^{(B-k)}} \right\}$$

$$+ \left\{ \sum_i^{\text{occ}} \frac{\left(\sum_r c_r^{(i)} c_{r'}^{(0')} \gamma_{rr'}^{(i)} \right)^2}{I_{Ai}^{(B)} - E_{Bo'}^{(A-i)}} + \sum_j^{\text{uno}} \frac{\left(\sum_r c_r^{(j)} c_{r'}^{(0')} \gamma_{rr'}^{(0')} \right)^2}{I_{Bo'}^{(A)} - E_{Aj}^{(B-o')}} \right\} \qquad (3.23)$$

where $0'$ denotes the singly occupied (SO) MO of B. Similarly, Eq. (3.21) becomes

$$D \sim \sum_i^{\text{occ}} \frac{c_r^{(i)\,2}}{\varepsilon_B - \varepsilon_{Ai}} \gamma_r^2 + \sum_j^{\text{uno}} \frac{c_r^{(j)\,2}}{\varepsilon_{Aj} - \varepsilon_{B'}} \gamma_r'^2 \qquad (3.24)$$

in which $\varepsilon_B = -E_{Bo'}^{(A-i)}$ and $\varepsilon_{B'} = -I_{Bo'}^{(A)}$. From the consideration of the form of γ_r and γ_r', it is worthy of note that, even in the interaction of a neutral molecule with a neutral radical, the local charge of atoms determines the magnitude of D. These equations are used for purposes which are similar to Eqs. (3.20) and (3.21).

In the case of degeneracy where one of the monotransferred configurations happens to have the same energy as the initial configuration, the first-order term of Eq. (3.10) appears. Obviously, such a case is possible only in regard to the transfer of one electron from HO MO of the donor molecule to LU MO of the acceptor molecule.

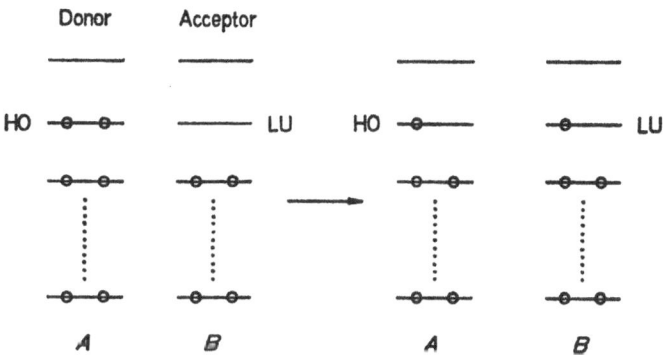

Fig. 3.3. The mode of donor-acceptor interaction

The equations corresponding to Eq. (3.14) and Eq. (3.20) are

$$D \sim |H_{0,\,HO \to LU} - S_{0,\,HO \to LU}\, H_{0,0}| \tag{3.25a}$$

$$D \sim \sqrt{2}\, |\sum_r c_r^{(HO)}\, c_{r'}^{(LU)}\, \gamma_{rr'}| \tag{3.25b}$$

In the special case of single-site overlapping as in Eq. (3.21), this becomes

$$D \sim \sqrt{2}\, c_{r'}^{(HO)}\, \gamma_r \qquad (\text{or } D \sim \sqrt{2}\, c_{r'}^{(LU)} \gamma_{r'}) \tag{3.26}$$

where

$$\gamma_r = c_{r'}^{(LU)}\, \gamma_{rr'} \qquad (\text{or } \gamma_{r'} = c_r^{(HO)}\, \gamma_{rr'}).$$

Eq. (3.25) stands for Mulliken's overlap and orientation principle. The charge-transfer interaction takes place according to the way in which the overlap of HO of the donor and LU of the acceptor becomes maximum. Particularly, the single-site interaction will occur at the position of the greatest HO density of the donor and at the position of the greatest LU density of the acceptor, as is seen from Eq. (3.26). In such cases the particular role of the frontier orbitals is evident.

Similar treatment has been made by Salem with discussions of many cases of special interest [112,113].

4. Principles Governing the Reaction Pathway

In the preceding section, the interaction energy between two reacting molecules has been discussed with the assumption of no nuclear configuration change. In the donor-acceptor interaction the delocalization stabilization is dominant. Eq. (3.25) indicates the importance of HO and LU in the donor-acceptor interaction. But the expression of Eq. (3.21) shows that in general cases the contribution of HO and LU to the quantity D is not so discriminative as those of the other MO's.

However, there exists a reason which makes the role of the frontier orbitals in the process of chemical reactions more essential than expected from the expression of D. This can be understood if the change in nuclear configuration along the reaction path is taken into consideration. The discussion of this point will be made with the aid of three principles governing the reaction pathway.

i) The principle of positional parallelism between charge transfer and bond interchange

The molecular orbital has, in general, its own nodal planes. The only MO which lacks nodal planes is the lowest-energy MO; all the other MO's must have at least one nodal plane in order to be orthogonal to the lowest-energy MO.

In view of the discussion in the preceding section, the *nodal property* of HO and LU is expected to be particularly important in the theory of chemical interaction. In reality, it has already been disclosed that the nodal property of the frontier orbitals plays an essential role in determining the orientation and steric course of electrocyclic reactions [44,49,56]. Schematic diagrams for the nodal property of π HO and LU of several conjugated molecules in the frame of LCAO MO scheme are indicated in Fig. 4.1, in which shaded and unshaded areas correspond to the positive and negative regions of MO's. In the following; we can understand that this property is significant in promoting alteration of the molecular shape in case of chemical interaction.

In common molecules, an atom is as a rule bonding with neighboring atoms in each occupied MO, and antibonding in each unoccupied MO. This circumstance is seen in every example illustrated in Fig. 4.1. Also

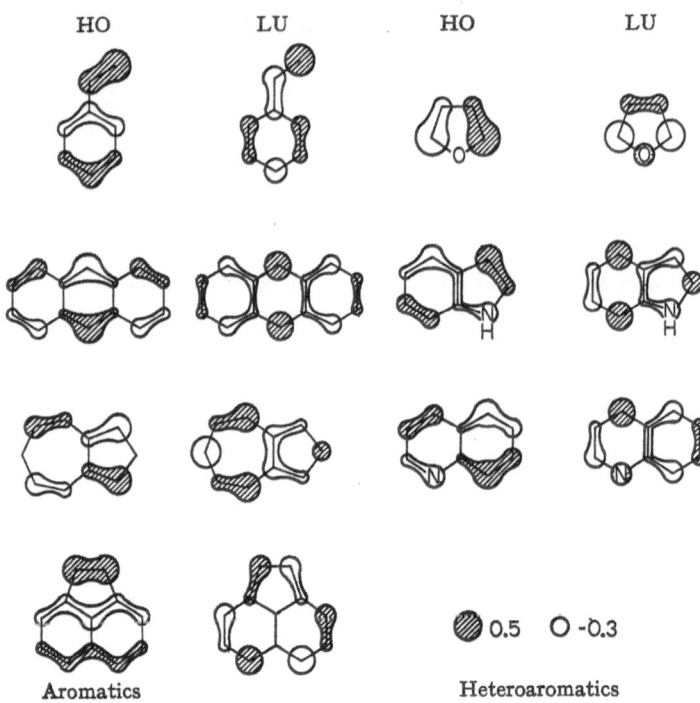

Aromatics   Heteroaromatics

●0.5   ○ -0.3

Fig. 4.1. The nodal property of π HO and LU of some conjugated molecules

the same is easily understood by investigating the simultaneous equations which are satisfied by the LCAO coefficients and the orbital energies. For instance, if we regard the Hückel MO (all overlap integrals neglected) for the $p\pi$ electrons of planar conjugated hydrocarbons, the following relations hold with respect to the ith MO:

$$(\alpha - \varepsilon_i)\,(c_r^{(i)})^2 + \sum_s^{\text{nei}} c_r^{(i)}\, c_s^{(i)}\, \beta = 0 \qquad (r = 1, 2, \text{-----}) \qquad (4.1)$$

in which α is the Coulomb integral of the rth carbon $2p\pi$ AO, β is the resonance integral between neighboring $2p\pi$ AO's, ε_i is the energy of the ith MO, $c_r^{(i)}$ is the LCAO coefficient of the rth AO in the ith MO, and $\sum\limits_s^{\text{nei}}$ means the summation over neighboring AO's of the rth AO. From Eq. (4.1),

$$\sum_s^{\text{nei}} c_r^{(i)}\, c_s^{(i)} = \frac{\alpha - \varepsilon_i}{(-\beta)}\,(c_r^{(i)})^2 \qquad (r = 1, 2, \text{-----}) \qquad (4.2)$$

are obtained. Since the usual hydrocarbons possess occupied MO's lower than α and unoccupied MO's higher than α, the quantity $\sum\limits_{s}^{\mathrm{nei}} c_r^{(i)} c_s^{(i)}$ is positive for an occupied MO and negative for an unoccupied MO and is proportional to the partial "π-electron density" at the rth atom in that MO. The quantity $\sum\limits_{s}^{\mathrm{nei}} c_r^{(i)} c_s^{(i)}$ represents the partial sum of bond orders of the rth atom with its neighbors.

Therefore, the position of the largest HO or LU density is at the same time the position where the bonds with neighboring atoms are as a whole most liable to loosening in case of electron-releasing or -accepting interaction, respectively. Since the HO or LU density is a measure of the ease of charge transfer interaction, as has been mentioned in the preceding section, this conclusion represents the parallelism between the charge transfer and the bond interchange in a molecule in chemical reactions. Namely, the charge transfer weakens the bonds with neighbors most at the position of the greatest frontier-orbital density.

Table 4.1. *The positional parallelism between* $(c_r^{(f)})^2$ *and* $\sum\limits_{s}^{\mathrm{nei}} c_s^{(f)}$ *in aromatic hydrocarbons by Pariser-Parr-Pople calculation* $[(f)$ *signifies* (HO) *or* (LU)$]$

Compound	position r	$(c_r^{(f)})^2$	$\pm \sum\limits_{s}^{\mathrm{nei}} c_r^{(f)} c_s^{(f)}$ [1]
Anthracene	9	0.19472	0.08735
	1	0.09512	0.03680
	2	0.04770	0.01966
Phenanthrene	9	0.16775	0.09051
	1	0.10550	0.07154
	3	0.09868	0.06598
	4	0.05843	0.03197
	2	0.00100	0.00034
Chrysene	6	0.14446	0.07687
	1	0.07978	0.04403
	4	0.06089	0.02985
	5	0.04871	0.02809
	3	0.04871	0.02168
	2	0.01427	0.00738

[1] $+$ sign for $(f) = $ (HO), and $-$ sign for $(f) = $ (LU).

Qualitatively, similar relationships are ascertained in *heteroaromatic systems* where the same conclusion is derived by a numerical calculation. In more elaborate calculations than the Hückel method, such as the Pariser-Parr-Pople approximation [21,23], similar distinct parallelisms are recognized [59] (Table 4.1). Essentially the same circumstances exist also

Table 4.2. *The positional parallelism between* $(c_r^{(LU)})^2$ *and* $v_r^{(LU)}$ *of hydrogens in 2-chlorobutane by the extended Hückel calculation*

Compound	Position r	$(c_r^{(LU)})^2$	$-v_r^{(LU)}$
	5	0.07887	0.05295
	3	0.06638	0.04564
	8	0.02119	0.01700
	1	0.00978	0.00661
	7	0.00158	0.00131
	6	0.00090	0.00041
	4	0.00044	0,00011
	2	0.00029	0.00006
	9	0.00000	0.00000

$(v_r^{(LU)} = 2 \sum_{s \neq r} c_r^{(LU)} c_s^{(LU)} s_{rs}; \; s_{rs}$: overlap integral)

in saturated compounds. This is assured [59] for instance by the extended Hückel calculation [31] (Table 4.2). Exemplifications by the various calculations mentioned above have indicated that the conclusion is independent of the level of approximation adopted, and is verified in a wide range of compounds.

ii) The principle of narrowing of inter-frontier level separation

It has been clarified that the charge-transfer interaction occurs at the position of the greatest frontier-orbital density which is simultaneously most susceptible to weakening of the bonds with the remaining part. This bond-weakening gives rise to a nuclear configuration change.

The direction of the nuclear configuration change is characterized by the mode of change in the energy level of HO MO of the donor and LU MO of the acceptor. The HO energy of the donor generally rises while the LU energy of the acceptor becomes lower in the event of charge transfer, since a bonding MO is made unstable by electron-releasing while an antibonding MO is stabilized by electron-accepting, in both cases through bond-weakening effectively, followed by a serious narrowing of inter-frontier energy level separation.

These circumstances become clear when we consider several common examples. The *Diels-Alder addition* of ethylene and butadiene is taken as the first and simplest example. Fig. 4.2a indicates the nodal property of HO and LU of ethylene and butadiene and the mode of charge transfer interaction. The ethylene HO is bonding while LU is antibonding. The

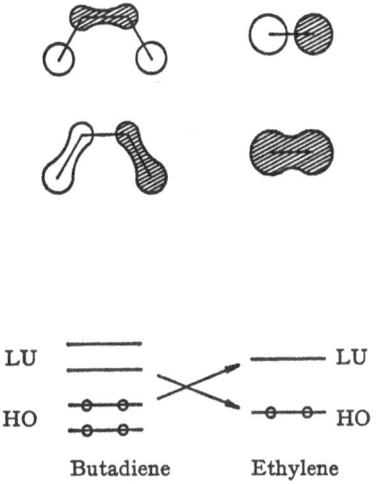

Fig. 4.2a. The nodal property of HO and LU in ethylene and butadiene

HO of butadiene is bonding in 1,2- and 3,4- π bonds and antibonding in the 2,3- π bonds, whereas the LU has the opposite bonding property. The charge transfer from HO of ethylene to LU of butadiene and that from HO of butadiene to LU of ethylene will both weaken the ethylene π bond and result in a double bond shift in butadiene. The change of bond lengths along the reaction path may reasonably be assumed by considering the direction of charge transfer and the nodal property of frontier orbitals. It is understood in Fig. 4.2b that the changes in frontier orbital energies are remarkable, in comparison with the other MO's, so that the inter-frontier separation becomes considerably narrower as the reaction proceeds. Such relations are commonly recognized with respect to many other dienes and dienophiles [59].

Similar results are obtained also in *sigma electron systems*. Various examples can be given in regard to the S_N2 reaction of a methyl halide with a halogen anion. E 2 reaction of alkyl halides, aromatic substitutions, solvation and desolvation, heterolytic addition to olefinic double bonds (see Fig. 4.3), and so on. In every reaction, the narrowing of inter-frontier energy level separation between the reactant and the reagent along the reaction path is verified by numerical calculation. This implies that the importance of the frontier orbitals is more than would be expected from the case in which these circumstances are not counted.

iii) The principle of growing frontier-electron density along the reaction path

The importance of the frontier-orbital AO coefficient is evident from Eqs. (3.21) and (3.26). The problem is how this quantity changes along the reaction path. It can be shown by actual calculation that the frontier-electron density generally increases as the reaction proceeds.

A typical example is given in the case of *aromatic substitutions*. The sum of the mobile bond orders of the bonds between the reaction center

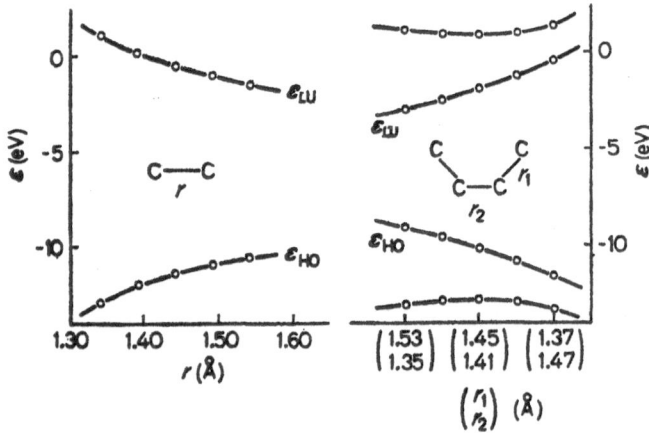

Fig. 4.2 b. The change in energy of MO's in ethylene and butadiene along the reaction path

and the neighboring atoms will gradually decrease according to the consideration stated in i) of this section, giving rise to the loosening of these bonds. The effect of this bond-loosening may be represented by a decrease in the absolute value of resonance integrals of these bonds, if the discussion is based e.g. on the Hückel MO approximation. What is to be made clear is whether or not the frontier-orbital density at the reaction center would in reality increase during the process of change which is represented schematically as the following:

$$\underset{I}{\beta \bigwedge \beta} \longrightarrow \underset{\substack{II \\ (\delta > 0)}}{\beta+\delta \bigwedge \beta+\delta} \longrightarrow \underset{\substack{III \\ (\Delta\beta < 0)}}{\Delta\beta \bigwedge \Delta\beta} \longrightarrow \underset{IV}{0 \quad 0}$$

in which β is the original value of resonance integral and r stands for the position of reaction. In actual reactions the change δ is rather small (Stage II). However, in order to illustrate the general tendency of the change, an extreme case where the increment is assumptively taken as $(-\beta)$ (Stage IV) may be considered. In that case the π AO at the reaction site is ultimately isolated. In Stage III, which is reached shortly before Stage IV, a small conjugation still remains between the π AO of the reaction center and the neighboring π AO's.

In order to understand qualitatively how the frontier-electron density, $(c_r^{(HO)})^2$ and $(c_r^{(LU)})^2$, as usual grows along the path (I)→(II) in planar conjugated hydrocarbons, it is convenient to take account of Stage III. In this stage it is easily proved that

$$\lim_{\Delta\beta \to 0} \{(c_r^{(HO)})^2 \quad \text{and} \quad (c_r^{(LU)})^2\} = \tfrac{1}{2} \tag{4.3}$$

provided that the hydrocarbon rest obtained by deleting the atom r from the original hydrocarbon molecule possesses one nonbonding MO, $\varepsilon = \alpha$. If the rest has n nonbonding MO's, $(r_r^{(HO)})^2$ and $(c_r^{(LU)})^2$ become $1/(n+1)$. Since the original frontier density values are in most cases far less than 0.5, Eq. (4.3) suggests the frontier-density growth along

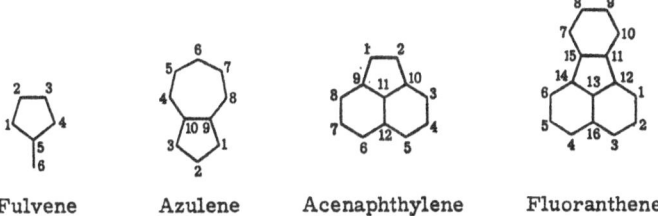

Fulvene   Azulene   Acenaphthylene   Fluoranthene

29

the reaction path. Eq. (4.3) is valid with respect to all so-called *"alternant" hydrocarbons*, and also in most of the actually reactive positions of nonalternant hydrocarbons, such as $(c_1^{(HO)})^2$ of fulvene, $(c_1^{(HO)})^2$ of acenaphthylene, and $(c_7^{(HO)})^2$ of fluoranthene.

In the case in which the hydrocarbon rest has no nonbonding MO, the discussion is rather complicated [59]. In several cases it holds that

$$\lim_{\Delta\beta\to 0} (c_r^{(HO)})^2 = 1 \qquad (4.4\,a)$$

$$\lim_{\Delta\beta\to 0} (c_r^{(LU)})^2 = 1 \qquad (4.4\,b)$$

Eq. (4.4a) is satisfied in the position 1 of azulene. Eq. (4.4b) is valid in position 6 of fulvene, position 6 of azulene, position 3 of fluoranthene, and position 5 of acenaphthylene. Even in a few exceptional cases where the previous relations do not hold, a consideration of the coulombic effect of attacking reagents leads to a conclusion favorable to the hypothesis of frontier density growth. An example of such cases is position 3 of

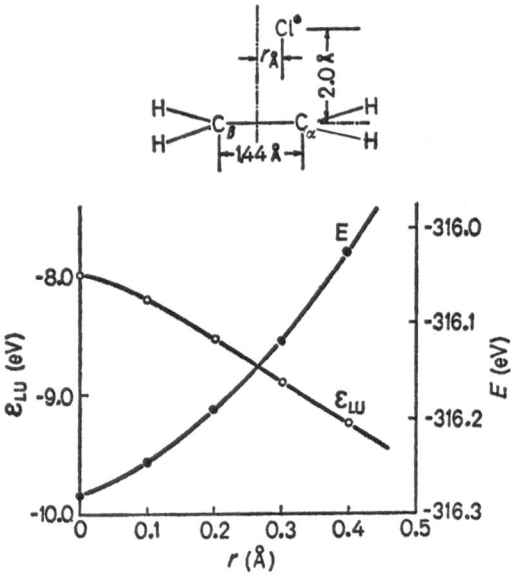

Fig. 4.3a. The change in the energy of LU, ε_{LU} and in the total energy, E, of ethylene-chlorine cation system

fluoranthene in HO MO. The rule of growing frontier density along the reaction path is essentially not violated by the adoption of more elaborate methods than the Hückel MO with respect to the calculation for aromatic substitutions.

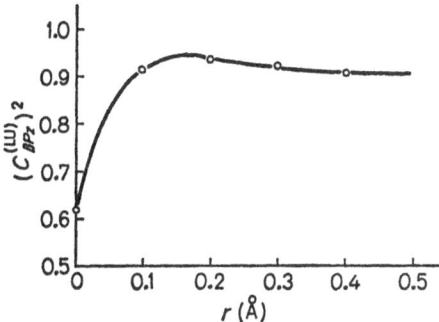

Fig. 4.3b. The changes in the LU partial population of p_z orbital at β-carbon $(c_{\beta p_z}^{(LU)})^2$, ethylene-chlorine cation system

The next example for this rule may be the *heterolytic addition of chlorine* to the C=C bond. Fig. 4.3b indicates the partial valence-inactive population [60] of the $2p_z$ AO of the β-carbon in LU, calculated by the extended Hückel method. It is seen that this quantity, $(c_{\beta p_z}^{(LU)})^2$, largely increases according to the approach of the chlorine cation to the carbon atom at which the addition is to take place, so that the reactivity of the β-position towards the second chlorine atom (anionic species) grows. Also Fig. 4.3a shows the decrease of the LU energy in the direction of the reaction path which has already been mentioned above.

5. General Orientation Rule

From the preceding discussions it is obvious that the three principles work co-operatively in promoting the reaction. As expected from Eq. (3.25), the charge-transfer interaction occurs dominantly at the position and in the direction in which the overlapping of HO and LU of the two reacting species becomes largest. The charge transfer brings about the local bond-weakening which is principally controlled by the nodal property of the frontier orbitals. The extent of the bond-loosening is positionally parallel to the frontier-orbital extension by the first principle. The weakening of bonds leads to the change in molecular shape in a definite direction, causing the narrowing of the inter-frontier energy level separation by way of the second principle, and simultaneously the frontier-electron density grows at the reaction center by the third principle. These effects will make the frontier term in the right side of Eq (3.14) more important, since the denominator becomes smaller and the numerator larger. Hence the contribution of the frontier-orbital interaction term to the delocalization part of the interaction energy (Eq. (3.14)) becomes larger, so that the amount of charge transfer increases, again in turn resulting in promotion of bond interchange near the reaction center, molecular shape deformation, narrowing of the frontier-level separation, and frontier-density growth. In this manner, the frontier-interaction term becomes more and more significant, leading to the approximate expression

$$D \sim \frac{|H_{0,\,\mathrm{HO}\to\mathrm{LU}} - S_{0,\,\mathrm{HO}\to\mathrm{LU}}\,H_{0,0}|^2}{H_{\mathrm{HO}\to\mathrm{LU},\,\mathrm{HO}\to\mathrm{LU}} - H_{0,0}} \tag{5.1}$$

even though the charge transfer in the initial stage of interaction is not so significant as in the obvious case of donor-acceptor interaction.

It should be noted here that the MO's which can take part in such a type of co-operation are evidently restricted to the particular MO's, HO and LU. The other MO's undergo only the minimum energy change which is absolutely required for the occurrence of reaction and may reasonably be assumed to be almost constant with regard to every possible reaction site of the same sort. This is understood from the following consideration. A stable molecule originally takes the nuclear configuration which is energetically most favorable. In the event of reaction, any change in nuclear configuration will bring about unstabilization. Such an unstabilization resembles the promotion in atoms in case of molecule formation. Accordingly, the change in molecular shape will occur in a direction which ensures the unstabilization is most powerfully eliminated. Any direction of change in which no energetic gain is expected will be avoided. The charge transfer between frontier orbitals gives rise to a change in molecular shape, which is thus automatically restricted to the neighbor of the reaction center in the reactant molecules. Such a *self-regulating nature* in the process of reaction will be the theoretical basis for the empirical rule which is known as "the principle of least motion" or *"the principle of least molecular deformation"* [61].

A chemical reaction is smoothly promoted by reducing the unstabilization energy ascribed to the change in molecular shape which is due to the interaction between reactant species. The most effective means of doing this is to give rise to a change by which the charge transfer between frontier MO's is effectuated. The charge transfer may be uni-directional or mutual according to the electron-donating or -accepting power of both reactants. All of the other directions of nuclear configuration change are rejected as bringing about little gain in stabilization energy.

It is thus evident that the reaction path is controlled by the frontier-orbital interaction. The position of reaction will be determined by the rule of maximum overlapping of frontier orbitals, that is, HO and LU MO's of the two reacting molecules. Sometimes SO takes the place of HO or LU in radicals or excited molecules. Hence, the general orientation principle would be as follows:

"A majority of chemical reactions are liable to take place at the position and in the direction where the overlapping of HO and LU of the respective reactants is maximum; in an electron-donating species, HO predominates in the overlapping interaction, whereas LU does so in an electron-accepting reactant; in the reacting species which have SO MO's, these play the part of HO or LU, or both."

Mention should be made here with respect to the intramolecular reactions. Some isomerization reactions, rearrangements, and the cyclization of a *conjugated olefinic chain* are the examples. The most dominant controlling factor in these cases seems to be the first-order interaction term [62,63], so that the HO—LU interaction is concealed. However, the same reaction can also be discussed by considering the frontier-orbital interaction between two parts of a molecule which are produced by a hypothetical division [64]. The HO—LU interaction has also been discussed with respect to the sigma- and pi-parts of conjugated molecules [56]. These two parts are regarded as if they were different molecules which are reacting with each other. A stereoselection rule which governs the reactions accompanying the hybridization change has been derived in this way. In this view, the particular MO's which seem to control the path of a chemical reaction, that is, HO, LU, and SO MO's, are referred to as *"generalized frontier orbitals"*.

The principle involved in the discussion mentioned above appears to be most general in nature, governing almost all kinds of chemical interaction, including intermolecular and intramolecular, as well as unicentric and multicentric. If the principle is applied to a unicentric reaction, it behaves as an orientation rule, and if it is employed to treat the multicentric reaction, as already mentioned in the discussion of Eq. (3.20), the stereoselection rule results [56,63,64].

It is to be noticed, however, that, considering cases like the *crystalfield* or *ligand-field interactions*, when the symmetry relationship between interacting MO's happens not to be favorable for the HO—LU interaction in a given „inflexible" configuration, the next-lying MO will temporarily act as the frontier orbital. Also in the case of *d*-orbital interaction, only the appropriate *d*-orbital which is symmetrically suitable for the interaction can play the part of the frontier orbital among the five degenerate, or almost degenerate, *d*-orbitals. The same will apply to cases of degenerate frontier orbitals (e.g. in benzene HO's and LU's) in general.

The general orientation rule described above is based solely on the consideration of charge-transfer interaction. Despite the discussions developed in Chap. 4, which may explain such a partiality to the charge-transfer term, the contribution of the other term to the interaction energy of Eq. (3.12) can never be completely disregarded. In particular, the Coulomb interaction term of Eq. (3.13) is frequently of importance. Klopman [109,110,111] took account of the effect of the first-order long-range Coulomb interaction term together with the second-order charge-transfer interaction for the purpose of discussing the chemical reactivity, introducing the concept of *"frontier-controlled"* and *"charge-controlled"* reactions. He states that to the former case belong the radical recombination and the reactions in the category of the *Woodward-Hoffmann rule* [51] as well as many conjugated hydrocarbon reactions.

6. Reactivity Indices

The reactivity index is the conventional theoretical quantity which is used as a measure of the relative rate of reactions of similar sort occurring in different positions in a molecule or in different molecules. As has already been mentioned in Chap. 2, most reactivity indices have been derived from LCAO MO calculations for unicentric reactions of planar π electron systems [65]. The theoretical indices for saturated molecules have also been put to use [50]. In the present section the discussion is limited to the indices derived from the theory developed in the preceding sections, since the other reactivity indices are presented in more detail than the frontier-electron theory in the usual textbooks [65,66] in this field.

The reactivity indices derived from the theory which has been developed in Chap. 3 are the frontier-electron density, the delocalizability, and the superdelocalizability, as has been mentioned in Chap. 2. These indices usually give predictions which are parallel with the general orientation rule mentioned in Chap. 5. The superdelocalizability is conventionally defined for the π-electron systems on the basis of Eq. (3.21) and Eq. (3.24) as a dimensionless quantity of a positive value by the following equations [49]:

i) For the reaction with an electrophilic reagent:

$$S_r^{(E)} = 2 \sum_i^{\text{occ}} \frac{c_r^{(i)2}}{\alpha - \varepsilon_i} (-\beta) \tag{6.1a}$$

Reactant   Reagent
                  (Electrophile)

ii) For the reaction with a nucleophilic reagent:

$$S_r^{(N)} = 2 \sum_i^{uno} \frac{c_r^{(i)\,2}}{\alpha - \varepsilon_i} (-\beta) \qquad (6.1\,b)$$

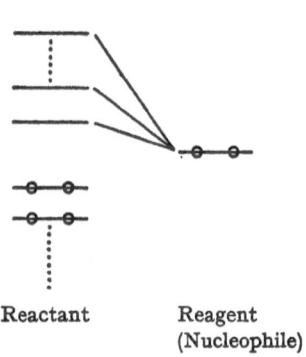

Reactant   Reagent
(Nucleophile)

iii) For the reaction with a radical reagent:

$$S_r^{(R)} = \sum_i^{occ} \frac{c_r^{(i)\,2}}{\alpha - \varepsilon_i} (-\beta) + \sum_i^{uno} \frac{c_r^{(i)\,2}}{\varepsilon_i - \alpha} (-\beta) \qquad (6.1\,c)$$

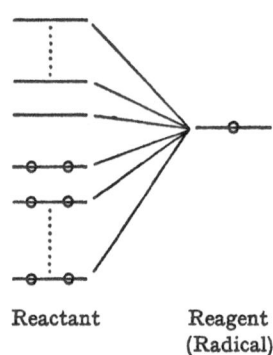

Reactant   Reagent
(Radical)

The Hückel integrals α and β are those which have appeared in Eq. (4.1). On inspecting the form of Eq. (6.1), the conventional character involved in the definition is obvious. First of all, the problem is the orbital energy of the reagent. In order to look into the orbital energy value of a reagent, it is necessary to take account of the solvation effect in condensed-phase reactions in general. In Eq. (6.1) the reagent MO energy is taken always equal to α. Such a convention would be allowable if the previous discussions on the nuclear configuration change along the reaction pathway are taken into account. Near the transition state, the MO energy is never equal to that of the initial reagent species. Rather the charge-transfer control will cause the levelling of HO and LU MO's as the reaction proceeds:

> a too high-lying LU level of an electrophile surrounded by solvent molecules will soon descend remarkably by desolvation, and, similarly, a too low HO level of a solvated nucleophile will be elevated by the same effect.
> Sometimes, a too low-lying LU level of a "bare" electrophile will immediately rise by the charge transfer from reactant molecules. Similar circumstances will appear in the case of a "bare" nucleophile.

For the purpose of comparing the reactivity towards *different* reagents, however, it may be more or less recommended to take into account the effect of the reagent orbital. In that case we need to go back to Eqs. (3.22) and (3.24). Such a type of modification of superdelocalizability has also been made [50,56,67].

The introduction of hybrid-based MO's into the theoretical treatment of paraffinic hydrocarbons [50] has made it possible to extend the applicability of Eqs. (3.21) and (3.24) to a wide variety of saturated compounds. The index has been called "delocalizability" of an atomic orbital and is defined by the following formulae [50,68]:

$$D_r^{(E)} = 2 \sum_i^{occ} \frac{c_i^{(i)\,2}}{\alpha' - \varepsilon_i} (-\beta') \tag{6.2a}$$

$$D_r^{(N)} = 2 \sum_i^{uno} \frac{c_r^{(i)\,2}}{\varepsilon_i - \alpha'} (-\beta') \tag{6.2b}$$

$$D_r^{(R)} = \sum_i^{occ} \frac{c_r^{(i)\,2}}{\alpha' - \varepsilon_i} (-\beta') + \sum_i^{uno} \frac{c_r^{(i)\,2}}{\varepsilon_i - \alpha'} (-\beta') \tag{6.2c}$$

in which $c_r^{(i)}$ is the coefficient of the rth AO in the ith MO of hybrid basis, ε_i is its energy, α' is the coulomb integral of an sp^3 hybrid in a carbon atom, and β' is the resonance integral between two sp^3 hybrids in a C—C bond [68]. Sometimes the standard quantities α' and β' are referred to the sp^2 hybridized state [50]. Obviously, the two treatments are in principle equivalent.

If the effect of the reagent orbital is wanted, the term α' in Eq. (6.2) may be replaced by the reagent MO energy, α_R [50].

In the light of the discussions made in Chap. 4, the contribution of the frontier term in the formula of S_r might be more important than expected from the expression. Such a consideration has early been made and a one-term approximation of S_r (denoted by S'_r) has been proposed [69]. Thus, S_r is approximated by the frontier term only:

$$S_r^{'(E)} = 2 \frac{c_r^{(HO)\,2}}{\alpha - \varepsilon_{HO}} (-\beta) \tag{6.3a}$$

$$S_r^{'(N)} = 2 \frac{c_r^{(LU)\,2}}{\varepsilon_{LU} - \alpha} (-\beta) \tag{6.3b}$$

$$S_r^{'(R)} = \frac{c_r^{(HO)\,2}}{\alpha - \varepsilon_{HO}} (-\beta) + \frac{c_r^{(LU)\,2}}{\varepsilon_{LU} - \alpha} (-\beta) \tag{6.3c}$$

Brown's reactivity index, Z-value [73], is also the one in which the frontier term solely controls the intramolecular orientation.

The contribution of the frontier orbitals would be maximized in certain special donor-acceptor reactions. The stabilization energy is represented by Eqs. (3.25) and (3.26). Even in a less extreme case, the frontier orbital contribution may be much more than in the expression of the superdelocalizability. If we adopt the approximation of Eq. (6.3), the intramolecular comparison of reactivity can be made only by the numerator value. In this way, it is understood that the frontier electron density, f_r, is qualified to be an intramolecular reactivity index. The finding of the parallelism between f_r and the experimental results has thus become the origin of the *frontier-electron theory*. The definition of f_r is hence as follows:

$$f_r^{(E)} = 2c_r^{(HO)^2} \tag{6.4a}$$

$$f_r^{(N)} = 2c_r^{(LU)^2} \tag{6.4b}$$

$$f_r^{(R)} = c_r^{(HO)^2} + c_r^{(LU)^2} \tag{6.4c}$$

In some cases half these values are adopted as f_r. The absolute value of the LCAO coefficient, $|c_r^{(t)}|$, serves as the measure of orbital extension, as well as the square value, $c_r^{(t)^2}$.

In the simple LCAO treatment in which the AO overlap is neglected, the "density" concept is rather clear-cut. An ambiguity arises in the case of inclusion of overlap. The extended Hückel calculation is one of the cases. The electron density is usually called "population" [70]. An analysis has been made with respect to the composition of population [71]. The population of the rth AO, q_r is defined by

$$q_r = 2 \sum_i^{occ} \sum_s c_r^{(t)} c_s^{(t)} s_{rs} \tag{6.5}$$

and is divided into two parts

$$q_r = p_r + v_r \tag{6.6}$$

where p_r is the "valence-inactive" part and v_r is the "valence-active" part, represented by

$$p_r = 2 \sum_i^{occ} c_r^{(t)^2} \tag{6.7}$$

$$v_r = 2 \sum_i^{occ} \sum_{s(r)} c_r^{(t)} c_s^{(t)} s_{rs} \tag{6.8}$$

in which s_{rs} is the overlap integral between the rth and sth AO's.

The value v_r is regarded as a measure of the extent to which the electron in the rth AO takes part in the bond formation with other atoms. In contrast with this, p_r is the part of population in the rth AO which is living there and responsible for the interaction with outside. Hence, in view of the role of the frontier orbital in the charge-transfer interaction, it is reasonable to take, as the frontier density, the *valence-inactive part* [72]. Namely,

$$f_r^{(E)} = p_r^{(HO)} = 2\, c_r^{(HO)^2} \tag{6.9a}$$

$$f_r^{(N)} = p_r^{(LU)} = 2\, c_r^{(LU)^2} \tag{6.9b}$$

$$f_r^{(R)} = \tfrac{1}{2}\left(p_r^{(HO)} + p_r^{(LU)}\right) = c_r^{(HO)^2} + c_r^{(LU)^2}. \tag{6.9c}$$

7. Various Examples

7.1. Qualitative Consideration of the HOMO-LUMO Interaction

The HO—LU interaction came early to the notice of theoreticians. Hückel [74] pointed out the role of LU in the alkaline reduction of naphthalene and anthracene. Moffitt [75] characterized the formation of SO_3, SO_2Cl_2, etc. by the reactions of SO_2 as an electron donor with the S-atom-localizing character of HO MO. Walsh [76] considered that the empirical result of producing nitro compounds in the reaction of the nitrite anion with the carbonium ion should be attributed to the HO of the NO_2 anion which is localized at the nitrogen atom.

Quite independently, of these fragmentary remarks, a distinctive role of HO (and later LU and SO, too) in unsaturated molecules was pointed out [43] in a general form and with substantiality (cf. Chap. 2). With respect to the molecular complex formation, the theory of charge-transfer force was proposed [47]. A clue to grasp the importance of HO—LU interaction was thus brought to light simultaneously both from the side of ionic reaction and from the side of molecular complex formation.

The Mulliken theory of overlap and orientation principle (cf. Chap. 2) predicts that stabilization in the molecular complex formation should essentially be determined by the overlap of the donor HO and the acceptor LU. The *iodine complex of trimethylamine* will take the form

$$\geqslant N ------ I - I$$

since the amine HO MO is the nitrogen lone-pair orbital and the LU of iodine is an antibonding $p\sigma$ orbital extending in the direction of the molecular axis. This is also consistent with experience.

The shape of the *complex of benzene and silver cation* is also explicable in a similar manner. The HO MO's of benzene are degenerate (e_{1g}) and have the symmetry as follows:

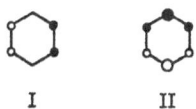

I   II

in which the size of circle symbolizes the orbital extension and the solid and hollow circles distinguish the different signs. Since the LU of silver cation ($5s$ AO) is obviously spherically symmetric, the location of silver cation on the symmetry axis of benzene will nullify the HO—LU overlapping. Hence, the cation is expected to lie above one of the C—C bonds such that

in conformity with experimental results [14]. A discussion on the silver cation complexes with various aromatic hydrocarbons has also been made [77].

A more complicated example has been discussed by Tsubomura [88]. The stability of the *quinhydrone-type complex* is ascribed to the symmetry

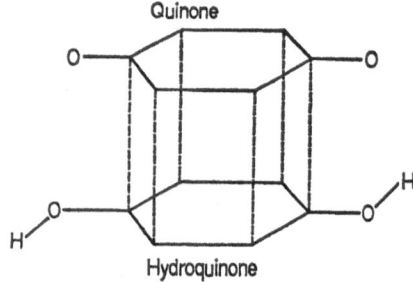

Fig. 7.1a. The quinhydrone complex

relation of HO of hydroquinone (b_1) and LU of quinone (also b_1) which is favourable for the HO—LU interaction. The orbital energy and symmetry relationship is indicated in Fig. 7.1a and b.

The same theory is useful for the understanding of the mode of orientation of ligands in many chelate compounds. The diagram [78] in

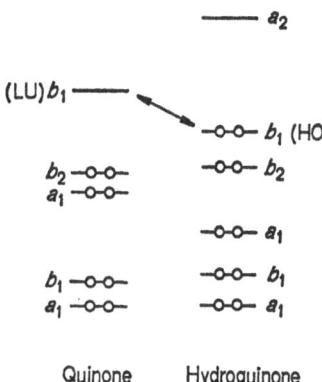

Quinone   Hydroquinone

Fig. 7.1b. The orbital symmetry relationship in quinone and hydroquinone

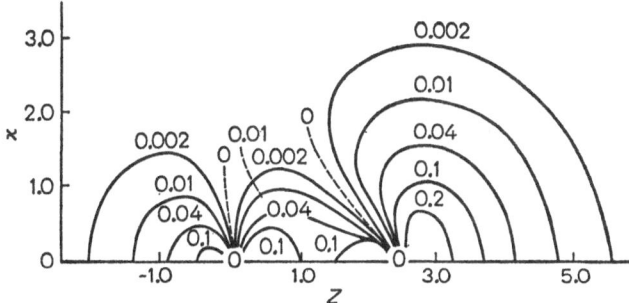

Fig. 7.2. The electron distribution diagram of HO (-0.55048 a.u.) of the CO molecule. (z: molecular axis)

Fig. 7.2 indicates the electron distribution of HO of carbon monoxide which largely localizes at the carbon atom [79]. This orbital resembles a lone-pair AO on the carbon atom and leads to the expectation that the carbon atom would behave as the electron-donating centre. As a matter of fact, the CO molecule coordinates with a metal cation by M—C—O type linkage (M represents a metal cation) in various metal carbonyl compounds. It is of interest to remark that the *total* electron population of the CO molecule has been shown by recent reliable calculation [80] to be rich on the oxygen atom in place of the carbon atom.

A similar result is obtained with respect to the cyanide anion CN⁻. The following mode of HO MO extension [81] underlies the M—C—N type orientation in chelate compounds:

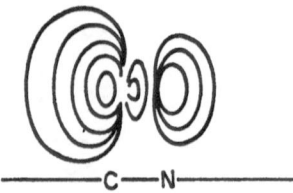

Fig. 7.3. The mode of extension of HO σ MO of CN⁻

A discussion along this line has been made in regard to the orientation of the hydrogen molecule in the *dissociative adsorption on metals* [82]. Thus, the interpretation of the function of *heterogeneous catalysis* on a molecular basis is no longer beyond our reach. The important role of LU MO in the process of polarographic reductions has also been discussed [83].

The antibonding LU MO of lithium hydride localizes more on the lithium atom than on the hydrogen atom, so that hydride anion will attack the lithium to form a linear anion.

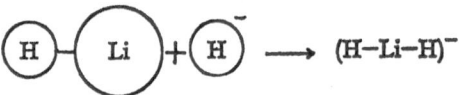

Recent calculation on pyridine shows that the HO MO is not the lone-pair orbital (σ HO) but a π orbital. Nevertheless, an acceptor-like proton attacks the σ HO instead of π MO.

	84)	85)	86) [1]
π HO (a_2)	-0.44725 a.u.	$I_p(\pi) = 9.28$ eV	$a_2 > b_1 > a_1$
π NHO (b_1)	-0.45856 a.u.		
σ HO (a_1)	-0.46543 a.u.	$I_p(n) = 10.54$ eV	

NHO: next-highest occupied (orbital)
I_p: ionization potential
[1] CNDO calculation

44

The reason that protonation takes place at the nitrogen lone-pair site, instead of nuclear protonation, is easily understood. In order to complete C-protonation, a large amount of energy is required for the hybridization change, whereas N-protonation does not need such energy. It is probable that a distant proton might approach the molecular plane along the extension of pi orbitals, entering then into the lone-pair region. The direction of σ protonation in pyridazine has also been discussed [87]. The result of calculation favours the configuration I.

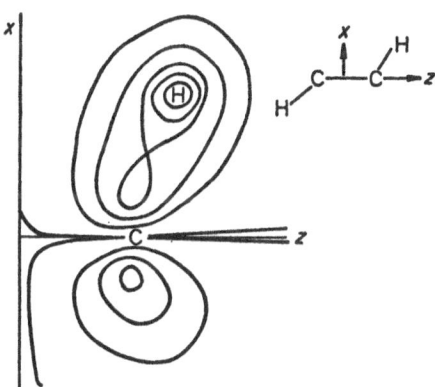

I   II

This reflects the maximum overlapping principle between HO and LU.

One of the suitable examples of sizable molecules may be *ethane*. The trans form belongs to the symmetry D_{3d}. The HO's are degenerate $1e_g$ MO, which are largely localized at C—H bonds and have bonding character on these bonds. The mode of extension is indicated below [89,90]. The LU is also localized at C—H bonds and antibonding. It is understood that most of the ionic and radical reactions of aliphatic hydrocarbons have some concern with the C—H bond.

Fig. 7.4. The HO of D_{3d} ethane. The electron distribution map in HCCH—plane

45

The HO of *cyclopropane* is degenerate $3e'$ MO [91]. The orbital I is responsible for a symmetric interaction, while orbital II is not. The protonation will take place in the σ plane as indicated. The mode of

Fig. 7.5. The HO MO's of cyclopropane

conjugation of the cyclopropane ring with an adjacent π electron system [91], is of interest from theoretical point of view. The NMR study of the *cyclopropylcarbonium ion* [92] favoured the orientation (a), which is easily interpreted by the interaction of σ HO of the cyclopropyl moiety (the above-mentioned orbital II) and π LU of the dimethyl carbinyl part ($2p$ orbital).

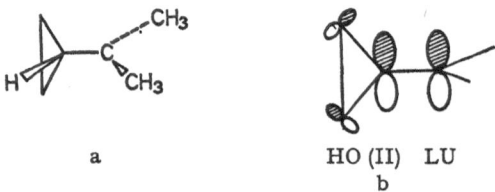

HO (II)   LU
b

Fig. 7.6a and b. The mode of conjugation of the cyclopropane ring with the adjacent π system

Many other studies gave results consistent with similar steric configurations [93,94]. The consequence of theoretical considerations also supports the conlcusion [95,96].

The configuration of dimers of BH_3, BR_3, AlH_3, AlX_3, AlR_3, etc. may be connected to the HO and LU extensions of monomers. Literature is available with regard to the knowledge of the HO and LU of *boron hydride* [97] and *aluminum hydride* and related compounds [98].

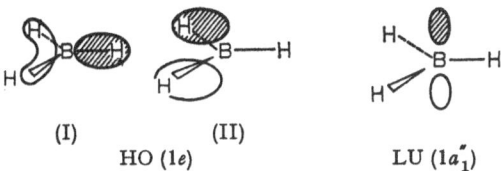

(I)   (II)

HO (1e)   LU (1a_1')

Fig. 7.7. The HO and LU of boron hydride

Among many examples of d-orbital interaction, only the following two are selected to illustrate the feature of HO—LU conjugation. One is the *cyclooctadiene-transition metal complex* [99]. The figure indicates the symmetry-favourable mode of interaction in a nickel complex. The electron configuration of nickel is $(3d)^8 (4s)^2$. The HO and LU of nickel can be provided from the partly occupied $3d$ shell from which symmetry-allowed occupied and unoccupied d orbitals for interaction with cyclo-octadiene orbitals are picked up.

The interaction of HO of cyclo-octadiene with unoccupied d orbital of nickel.

The interaction of LU of cyclo-octadiene with occupied d orbital of nickel.

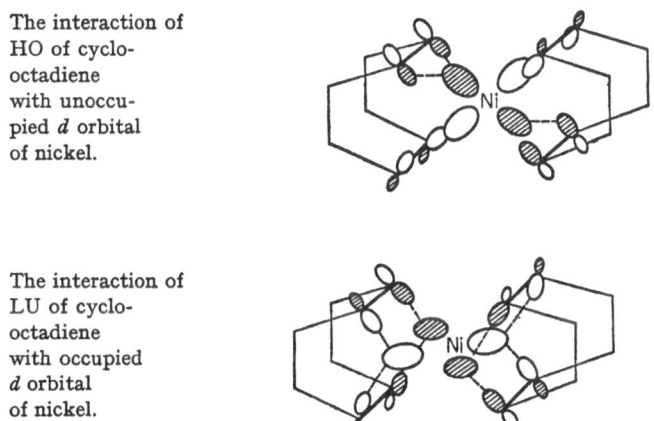

Fig. 7.8. The mode of orbital interaction in Ni-cyclooctadiene chelate

Similar chelate compounds are known, like [100]

47

The multifarious character of d-orbital symmetry provides a possibility of explaining the catalytic action of transition metal compounds. One example is the catalytic disproportionation of olefins [101].

$$2\,R-CH{=}CH-R' \underset{WCl_6}{\overset{\longrightarrow}{\longleftarrow}} R-CH{=}CH-R \;+\; R'-CH-CH-R'$$
$$(1{:}1)$$

It is probable that the tungsten d-orbital might facilitate a square-form interaction of two olefinic bonds.

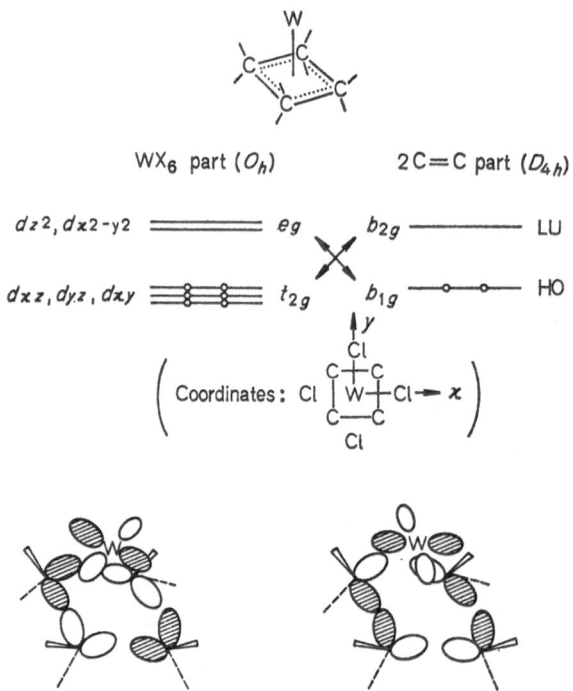

The occupied d_{xy} of W in interaction with the LU of 2 C=C system

The unoccupied $d_{x^2-y^2}$ of W in interaction with the HO of 2 C=C system

Fig. 7.9. The possible mode of orbital overlapping of tungsten d orbitals with two ethylenic bonds

The consideration of HO–LU interaction is useful also in the interpretation of the stability of "nonclassical" carbonium ions. For instance, the *7-norbornenyl cation* would be stabilized by the symmetry-allowed interaction of LU of 7 methine and HO of olefinic part [155], whereas the same anion would get no such stabilization on account of symmetry prohibition.

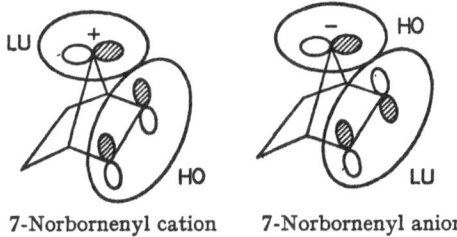

7-Norbornenyl cation   7-Norbornenyl anion

Fig. 7.10. The conjugation stabilization in 7-norbornenyl ions

The CNDO calculation gives a result of the same trend [102].

The stability of *benzvalene* may be discussed by dividing the molecule into two parts, the tetramethine part and the dimethine part, as illustrated below.

The favourable relationship of orbital symmetry will contribute the delocalization stabilization. Such a consideration by "partition technique" is frequently useful.

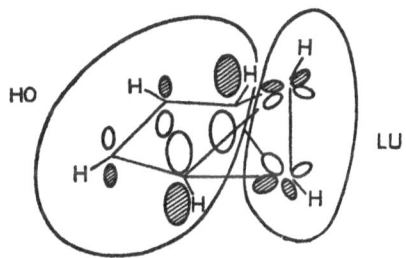

Fig. 7.11. The mode of stabilization in benzvalene

7.2. The Role of SO MO's

As has been mentioned in Chap. 5, the singly occupied MO in odd-electron molecules and radicals plays the role of HO or LU or both MO's according to the orbital energy relationship and the orbital overlapping situation. The importance of SO distribution is easily understood

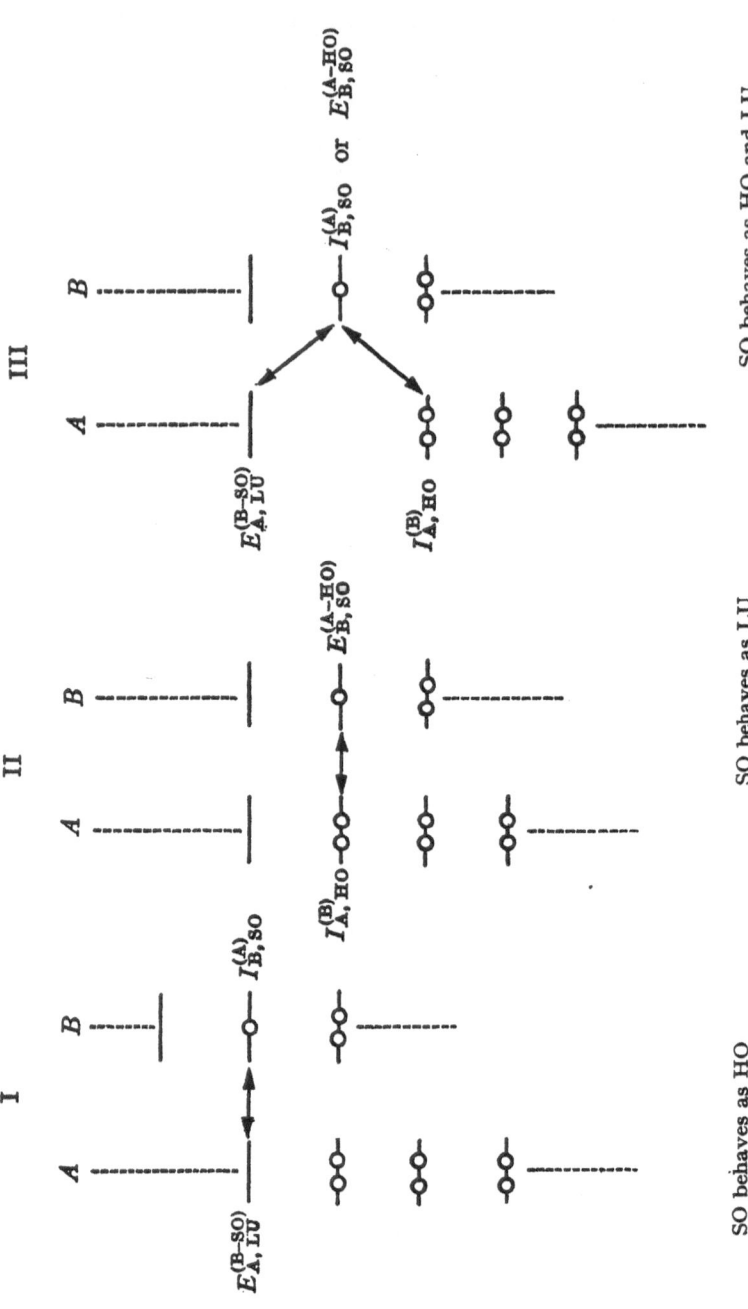

Fig. 7.12. The behavior of SO MO in interaction with a ground-state molecule[a] ($I_{B,SO}^{(A)}$, $E_{A,LU}^{(B-SO)}$, etc., denote the quantities mentioned in Chap. 3)

[a] This diagram is written in the sense of the "restricted Hartree-Fock" scheme[18]. In the "unrestricted Hartree-Fock"[19] sense each orbital of radical B is "singly" occupied and LU is higher and HO is lower than the restricted Hartree-Fock SO, respectively (cf. Chap. 1)

by reference to Eq. (3.23) in which the second bracket term in the right side will make a large contribution (see Fig. 7.12). Notice that even in case III in Fig. 7.12 the "mutual" charge-transfer from SO of B to LU of A and from HO of A to SO of B is of particular significance in the sense that has been mentioned in Chap. 4.

In the methyl radical, the reaction takes place in the direction of SO ($2p\pi$ of central carbon) extension, that is to say, the direction perpendicular to the molecular plane. Walsh [76] correlated the remarkable localization of SO at the nitrogen atom in NO_2 to the experimental results indicating that NO_2 abstracts hydrogen from other molecules to form HNO_2 rather than HONO, combines with NO to form $ON-NO_2$, dimerizes to produce O_2N-NO_2, and so forth. Also he pointed out that the SO MO of ClCO is highly localized at the carbon atom, which is connected with the production of Cl_2CO in the reaction with Cl_2. The SO extension of NO_2 is schematically shown below [103].

According to the recently elaborated calculation on BeH using 50-configuration wave function [104], which gives the value of -15.221 a. u. for the $^2\sum^+$ ground-state energy in comparison with the experimental

values of -15.254 a. u., the SO (3σ) MO of the BeH radical is largely extended in Be to the outside direction, which suggests the linear form of BeH_2 molecule (H—Be—H).

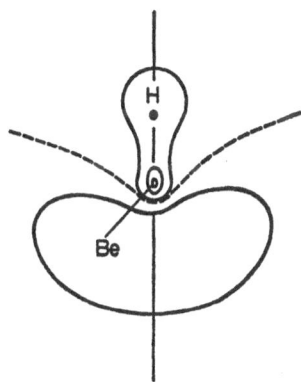

Fig. 7.13. The extension of SO MO of BeH

As for more complicated molecules, the exo-addition [106] in the 2-norbonyl radical was explained from the point of view of SO extension [105]. The 2-carbon is not exactly sp^2 hybridized but extends more in the exo direction than in the endo direction [156]. The nonplanarity of almost-sp^2 carbon in radicals is also expected in 2-chloroethyl and 4-t-butylcyclo-hexyl in which stereoselective recombinations are known. A rather ex-aggerated illustration of the mode of extension of SO MO is given below [107].

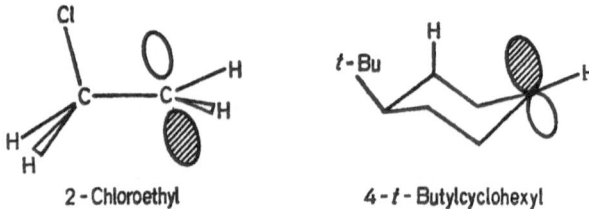

2 - Chloroethyl   4 - t - Butylcyclohexyl

Fig. 7.14. The mode of SO extension in 2-chloroethyl and 4-t-butylcyclohexyl

The SO MO's in excited states behave in a way similar to those in radicals. Walsh [76] noticed that both of the SO MO's in the first excited state of the SO_2 molecule localize largely at the *sulphur atom*. This was correlated to the formation of bonds at the sulphur atom in the photo-chemical reaction of SO_2:

$$S\begin{matrix} \diagup O \\ \diagdown O \end{matrix} + O_2 \longrightarrow O-S\begin{matrix} \diagup O \\ \diagdown O \end{matrix}$$

$$S\begin{matrix} \diagup O \\ \diagdown O \end{matrix} + Cl_2 \longrightarrow \begin{matrix} Cl \\ Cl \end{matrix}\!\!\diagdown\!\!S\!\!\diagup\begin{matrix} O \\ O \end{matrix}$$

The theoretical significance of SO MO's in the excited-state molecules was discussed in detail [108,62]. One of these SO's, or both, play important parts in excited-state reactions.

Even in reactions involving excited states or in reactions between two radicals, the primary interaction which determines the reactivity is thought to proceed adiabatically. The probability of nonadiabatic charge transfer also may not be ignored between a molecular specie with small ionization potential and a specie with large electron affinity, in particular in the form of free, gaseous, or nonsolvated state. In that

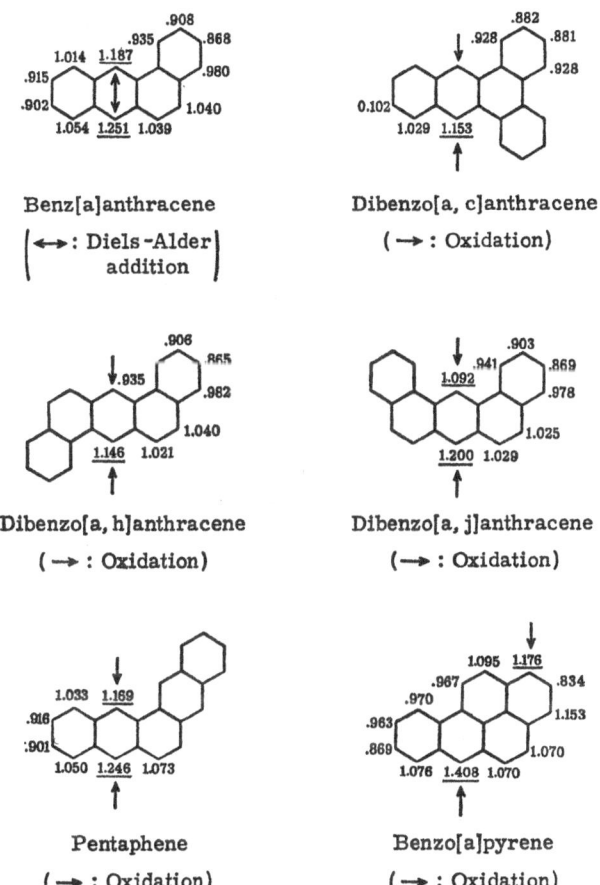

Fig. 7.15. The S_r values for aromatic hydrocarbons. (Positions of reaction are denoted by an arrow [135])

case, a zeroth order perturbation term which does not depend on the reaction position in the molecule will appear in the right side of Eq. (3.12). However, the orientation principle is not affected.

7.3. Aromatic Substitutions and Additions

As has already been mentioned in Chap. 2, aromatic substitution was the first object of theoretical treatment of chemical reactivity. The reactivity indices of Chap. 6 have also been first applied to the aromatic substitution. Since existing papers [43] and reviews [44,65] are available for the purpose of verifying the usefulness of the indices, f_r and S_r, only a few supplementary remarks are added here.

Fig. 7.15 is constructed from reactivity diagrams of aromatic hydrocarbons already published. The reactivity of fluoranthene has often been investigated in detail from both, the experimental and theoretical aspects [116]. The values of $f_r^{(E)}$ calculated by the Pariser-Parr method (SCF) [117] as well as by the Hückel MO (HMO) modified by considering

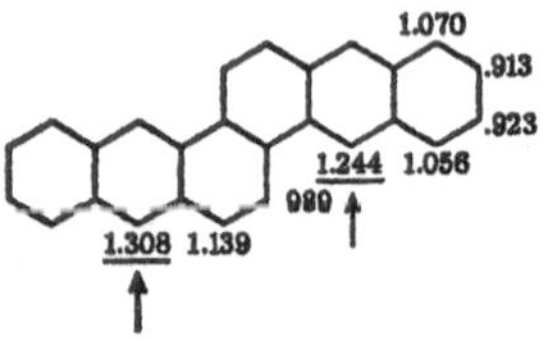

Dibenzo[b, k]chrysene

(→ : Oxidation)

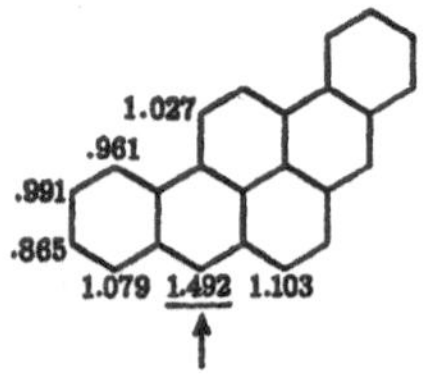

Benzo[rst]pentaphene

(→ : -CHO)

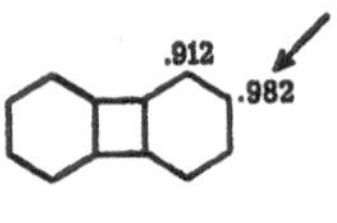

Biphenylene

(→ : Various reactions)

Fig. 7.15 (continued)

the next occupied MO, which lies very close to the HO, according to the appropriate procedure described in literature reference 9b give correctly the experimental order of reactivity $3>8>7>1>2$. The value $S_r^{(E)}$, also based on the Pariser-Parr calculation and with α put equal to the mean value of HO and LU MO energies, shows the order $3>7>8>2>1$ in slight disagreement with experiment.

These indices were initially used in the frame of Hückel MO method. But the theory has been shown to be valid also in more elaborate methods

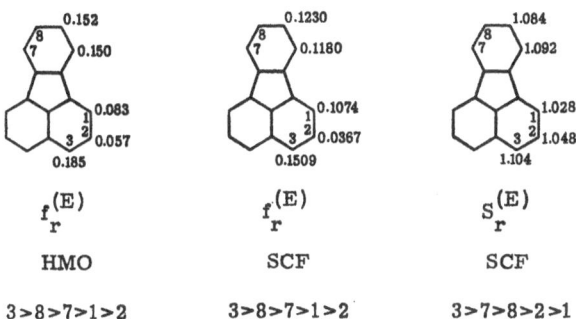

$f_r^{(E)}$	$f_r^{(E)}$	$S_r^{(E)}$
HMO	SCF	SCF
$3>8>7>1>2$	$3>8>7>1>2$	$3>7>8>2>1$

Fig. 7.16. The reactivity of fluoranthene

of calculation. Such an "approximation-invariant" character of the theory has already been discussed [44]. One of the recent examples is pyrrole. Clementi's very accurate calculation [114] gives no different result with respect to the inference of the reactive position (Fig. 7.17).

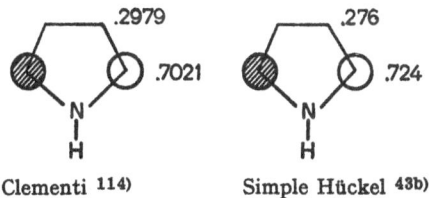

Clementi [114]   Simple Hückel [43b]

Fig. 7.17. The frontier electron density $f_r^{(E)}$ for pyrrole

A ten π electron heterocycle, *imidazo [1,2-α] pyridine* was studied by Paudler and Blewitt [115]. The protonation occurred at N_1, which was calculated to have a total π electron density less than N_4 (Fig.

7.18a). They calculated $f_r^{(E)}$ distribution to find that this is larger at N_1 than N_4 (Fig. 7.18b). Bromination took place at C_3 where both q_r and $f_r^{(E)}$ are largest.

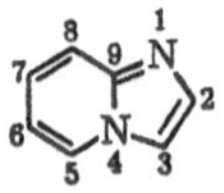

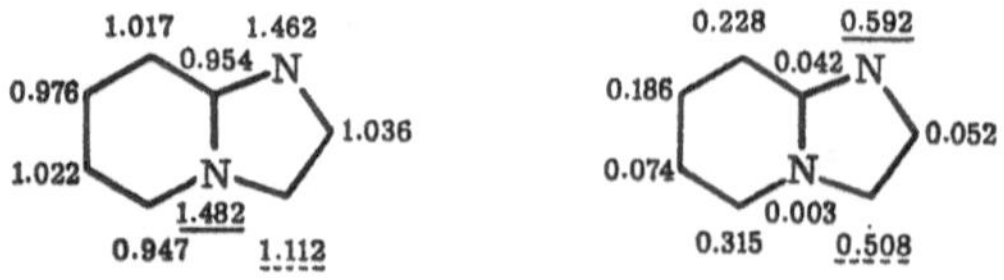

a   b

Fig. 7.18a and b. The total π electron density, q_r, and $f_r^{(E)}$ in imidazo[1,2-α]pyridine [115]. a) q_r, b) $f_r^{(E)}$

One example showing a serious "discrepancy" of the frontier electron method was reported by Dewar [118,119]. This is *10,9-borazaphenanthrene*, and the value of $f_r^{(E)}$ was reported to have been calculated by the Pople method, but the parameters used were not indicated. Fujimoto's calculation by the Pariser-Parr-Pople method [120], in perfect disagreement with Dewar's, gives the most reactive position as 8, which parallels experiment. The ambiguity involved in the integral values adopted seems to be serious, so that the establishment of parametrization for boron heterocycles is desirable.

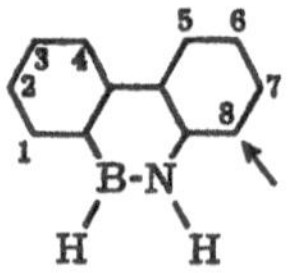

A comprehensive study has been made by the use of S_r with respect to the antioxydant action of amine compounds [134]. Several beautiful parallelisms are found between the activity and the superdelocalizability.

7.4 Reactivity of Hydrogens in Saturated Compounds

The reactivity of hydrogens at various positions of aliphatic and alicyclic hydrocarbons and their derivatives in various reactions is successfully interpreted by the theoretical indices, D_r and f_r, mentioned in Chap. 6. Most of the results obtained were reviewed in reference 16 and are not repeated here.

The HO and LU MO of propane are available from the result of calculation by Katagiri and Sandorfy [29] which is based on the method already mentioned in Chap. 1. Fig. 7.19 indicates the result. Both HO and LU localize more at secondary CH bonds than at primary CH bonds, reflecting the reactivity of C_3H_8.

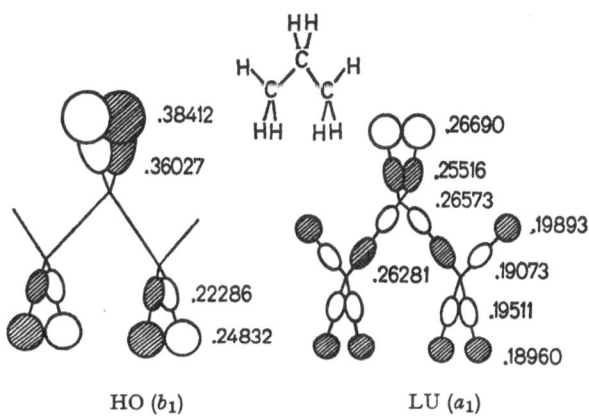

Fig. 7.19. The hybrid-based MO coefficients (absolute value) in propane. [Shaded and unshaded areas correspond to different signs of AO coefficients (+ lobe and − lobe)]

The reactivity of hydrogens in norbornane towards abstraction is of interest since the difference between two hydrogen atoms attached to the same carbon atom of position 2 can well be explained. The frontier electron density values [105] are in accord with the reactive *exo* hydrogen (Fig. 7.20).

Adamantane-type cage hydrocarbons have become a new target of theoretical investigation. The tertiary hydrogens which are known to be

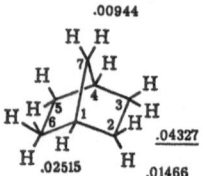

Fig. 7.20. The (HO + LU) density values of hydrogen atoms in norbornane

reactive towards homolytic are shown to have larger $D_r^{(R)}$ values than secondary ones (Fig. 7.21) [121].

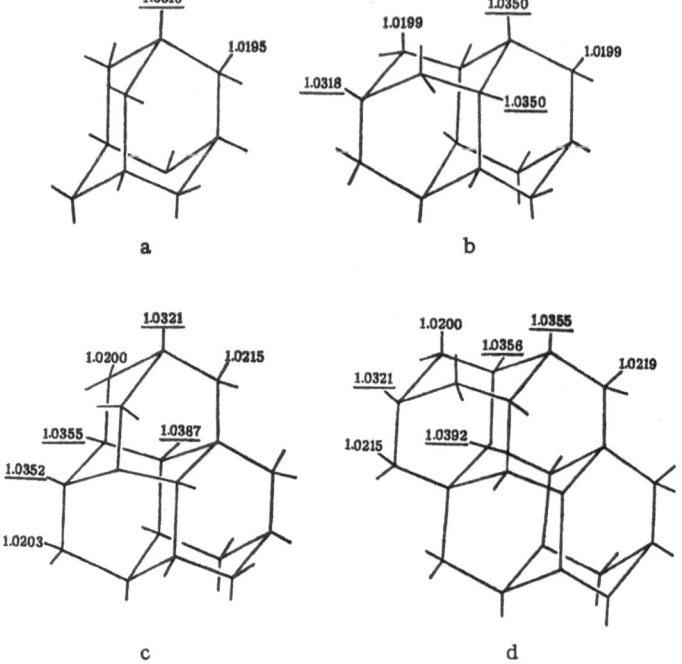

Fig. 7.21 a—d. The $D_r^{(R)}$ values of hydrogens in adamantanes. a) Adamantane, b) Diamantane, c) Triamantane, d) Tetramantane

The important role of LU MO in the nucleophilic reactions of saturated hydrocarbons bearing nucleophilic substituents (halogens, alkoxy⁻, acyloxy⁻, RSO₂O⁻, etc.) in the molecule has been pointed out [122,123].

The LU MO of ethyl chloride (*trans* form) extends in the region of the α carbon to the direction opposite the side of the chlorine atom and also in the region of the hydrogen atom *trans* coplanar to the chlorine atom [124]. The former is responsible for the attack of nucleophile in S_N2 reactions, and the latter for the attack in E 2 reactions.

The value of $f_r^{(N)}$ has been calculated with respect to various *halogenoparaffins* [122,123,125]. Only one example is mentioned here. The LU density on hydrogen atoms in *t*-amyl chloride is indicated in Fig. 7.22. This MO highly localizes on *trans* hydrogens, and the hydrogen atom on C_3 has greater density than the hydrogen atom on C_1, corresponding to the reactivity of *trans* elimination and the Saytzeff rule.

Fig. 7.23 shows the example of 2-exo-chloronorbornane [123] which suggests the occurrence of the *exo-cis* elimination in conformity with experiment [126].

The S_N2 and E 2 reactions usually take place more or less concurrently.

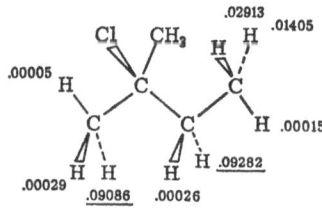

Fig. 7.22. the hydrogen $(C_r^{(LU)})^2$ values of *t*-amyl chloride

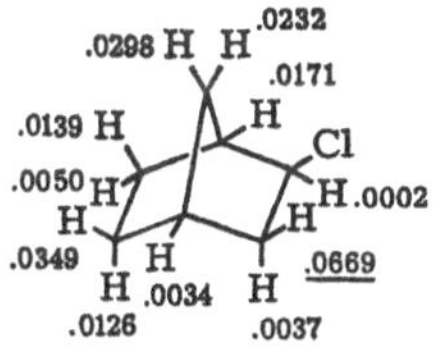

Fig. 7.23. The hydrogen $(C_r^{(LU)})^2$ values of 2-*exo*-chloronorboranane

The order of reactivity in the series of RBr is known as

$$S_N2: \quad CH_3 > C_2H_5 > (CH_3)_2CH > (CH_3)_3C$$

$$E\ 2: \quad C_2H_5 < (CH_3)_2CH < (CH_3)_3C$$

which are successfully interpreted by the orbital coefficients in LU [125]. Also the base-catalyzed hydrolysis of carboxylic esters with acyl-oxygen fission can be treated in a similar fashion [125]. The LU density of protonated ketones explains the reactivity of ketones in acid-catalyzed halogenation [125].

The reaction of S_N2', that is, the *bimolecular nucleophilic substitution with allyl rearrangement*

$$\underset{|}{\overset{|}{C}}-\overset{|}{C}=\overset{|}{C}-X \quad \xrightarrow{B^-} \quad B-\overset{|}{\underset{|}{C}}-\overset{|}{C}=\overset{|}{C} \quad + \quad X^-$$

is known to occur in the direction *cis* to the leaving nucleophilic group [127,128]. The LU MO of allyl chloride extends more in the direction *cis* to the chlorine atom than in the direction *trans* at the γ carbon atom [129]. The opening of the epoxy ring by the hydride anion is known to take place in the direction *trans* to the oxygen atom [130].

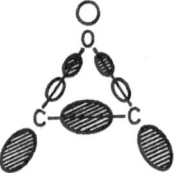

The extension of LU MO [129] explains the direction of attack of H⁻. The strong antibonding character of the C—O bond corresponds to the ring-opening reactivity.

Fig. 7.24. The LU MO of ethylene oxide

The base-catalyzed allyl rearrangement of olefins can be treated by the LU orbital density criteria [132]. The LU orbital remarkably localizes at the hydrogen atoms attached to the β carbon to the double bond in various olefins, as is shown in Fig. 7.25 by the use of a few examples, and is in conformity with the experimental fact that the β hydrogen is first abstracted by the base.

The β hydrogen atoms are as a whole antibonding with the remaining part in LU MO, so that the charge-transfer to LU from the base easily comes to release these hydrogens. Similar double-bond shift reactions have also been treated [133].

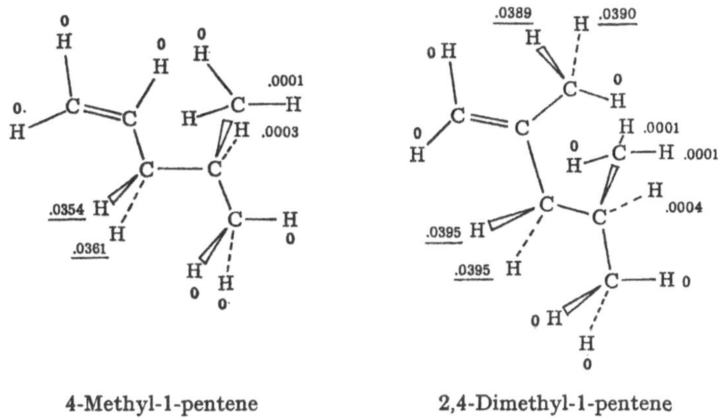

4-Methyl-1-pentene   2,4-Dimethyl-1-pentene

Fig. 7.25. The value of $2(C_r^{(LU)})^2$ in olefins

7.5 Stereoselective Reactions

In the reactions mentioned in the preceding sections, several "stereoselective" processes have been involved. Various examples have verified that the extension of singly-occupied MO determines the favorable spatial direction of interaction with other species. If there are two such nonequivalent directions in the molecule, the reaction will become stereoselective. Two or more hydrogen atoms attached to the same carbon atom are in some cases nonequivalent. Such a nonequivalence becomes a cause of stereoselectivity and has been explained theoretically. Also several cases have been mentioned in which some nucleophiles selectively attack the molecule from a certain spatial direction.

The general relation which must be satisfied in order to bring about an appreciable stabilization energy in the chemical interaction has been given by Eq. (3.20) and Eq. (3.25b). Such relations frequently provide a "selection rule" for the occurrence of stereoselective reactions.

Such a selection rule was first found in the *Diels-Alder addition* [44]. Eq. (3.25b) is simply applied to the interaction between the HO of dienes and the LU of dienophiles, obtaining

$$D \sim \sqrt{2} \mid c_r^{(\text{HO})} c_{r'}^{(\text{LU})} + c_s^{(\text{HO})} c_{s'}^{(\text{LU})} \mid \cdot \mid \gamma \mid \tag{7.1}$$

where $\gamma_{rr'}$ and $\gamma_{ss'}$ are taken to be equal, and r and s denote the 1,4-positions of the diene, and r' and s' the corresponding 1,2-positions of the dienophile (Fig. 7.26).

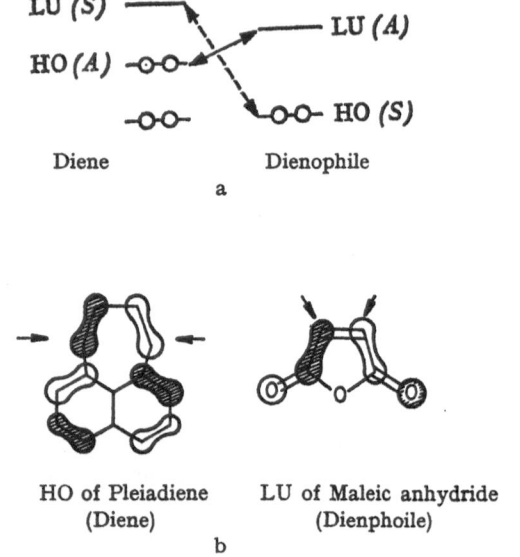

Fig. 7.26a and b. The diene-dienophile interaction. a) Orbital relationship, b) An example. (S: symmetric, A: antisymmetric)

It is easily understood from Eq. (7.1) that the signs of $\{c_r^{(\text{HO})} c_{r'}^{(\text{LU})}\}$ and $\{c_s^{(\text{HO})} c_{s'}^{(\text{LU})}\}$ are required to be the same in order for D to have an appreciable magnitude. All the examples of combination of diene and dienophile in which reaction actually takes place were found to satisfy this condition [44]. It is to be noticed that this conclusion is independent on the sign of each AO adopted (Return to Eq. (3.20)).

Similar relationships have been established with regard to the *1,3-dipolar addition* and the *photodimerization of olefins* [62]. The HO—LU

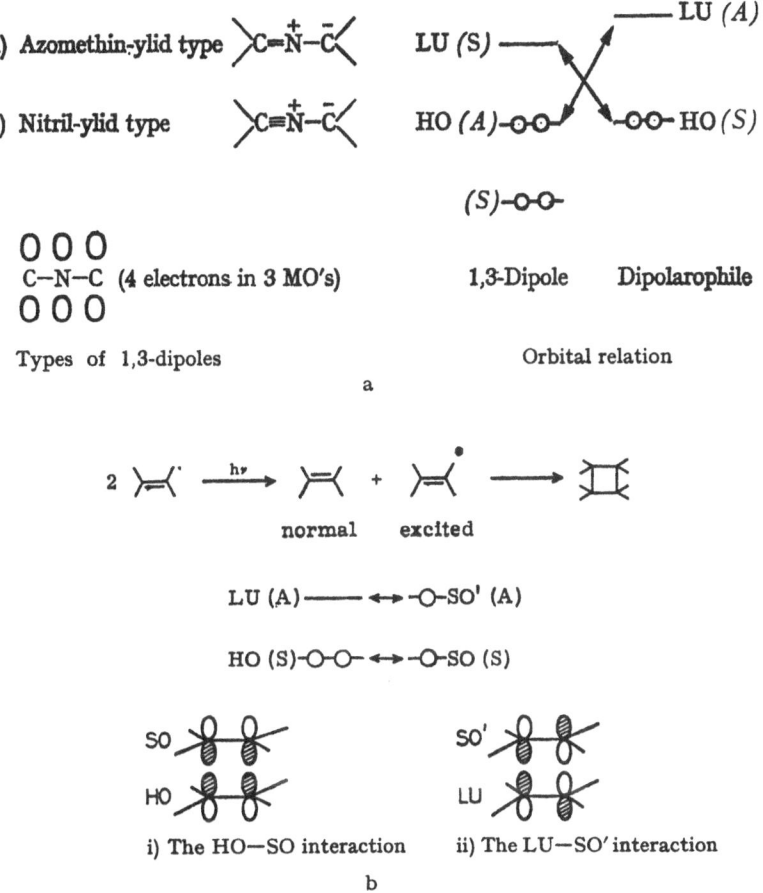

Fig. 7.27a and b. Orbital relationships in 1,3-dipolar additions and photodimerization of olefins. a) 1,3-Dipolar addition, b) Photodimerization of olefins

symmetry relationship is favourable for the overlap stabilization, as is seen in Fig. 7.27. The essential part of 1,3-dipoles is a four-π electron system with three π AO's, and the orbital symmetry relation is in favour of the interaction of 1 and 3 positions of 1,3-dipole with 1 and 2 positions of dipolarophile, respectively. The protodimerization of olefins, in which one reactant is thought to be an excited molecule, may proceed by way of a similar favourable orbital symmetry relationship. Other cyclic dimerizations and cyclic rearrangements like Claisen and Cope types were similarly treated [62].

Theoretical considerations in the same fashion enable predication of the possible configuration of the transition state. Eq. (3.25b) for the multicentre interaction is utilized. Hoffmann and Woodward [136] used such methods to explain the endo-exo selectivity of the Diels-Alder reaction (Fig. 7.28). The maximum overlapping criteria of the Alder rule is in this case valid. The prevalence of the endo-addition is experimentally known [137].

Similar discussion is possible with respect to the transition state of the *Claisen* and *Cope rearrangements* [138]. These can be treated similarly. Fig. 7.29a indicates that the symmetry of SO MO suggests *cis-cis* interaction with the six-membered structure for the transition state, but the chair-boat selectivity is not determined by the SO—SO interaction. The overlapping of LU' and HO' plays a secondary role. Fig. 7.29 shows that the boat form is unfavourable in comparison with chair form on account of the different signs of LU' and HO' at the central carbons. Similar consideration is possible with respect to the extended Cope rearrangement (Fig. 7.29.b). The predominance of the chair-form transition state is known both in the Claisen [139] and the Cope rearrangements [140].

The theory of stereoselectivity found in intramolecular hydrogen migration in *polyenes* was disclosed by Woodward and Hoffmann [51]. The HO—LU interaction criterion is very conveniently applied to this problem [64]. The LU of the C—H sigma part participates in interaction with the HO of the polyene π part. The mode of explanation is clear-cut

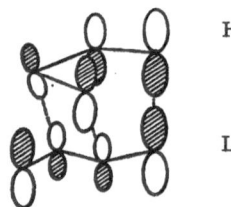

HO of diene

LU of dienophile

Endo-configuration

Fig. 7.28. The exo-endo selectivity in Diels-Alder reactions

in Fig. 7.30. For instance, in case a), when the carbon hybrid at the C—H σ part is given the same sign as the end π AO of the butadiene part, the

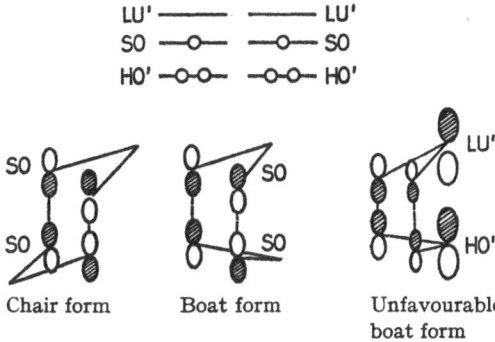

Chair form   Boat form   Unfavourable
boat form

Fig. 7.29a. The transition state in the Cope rearrangements. Cope rearrangement

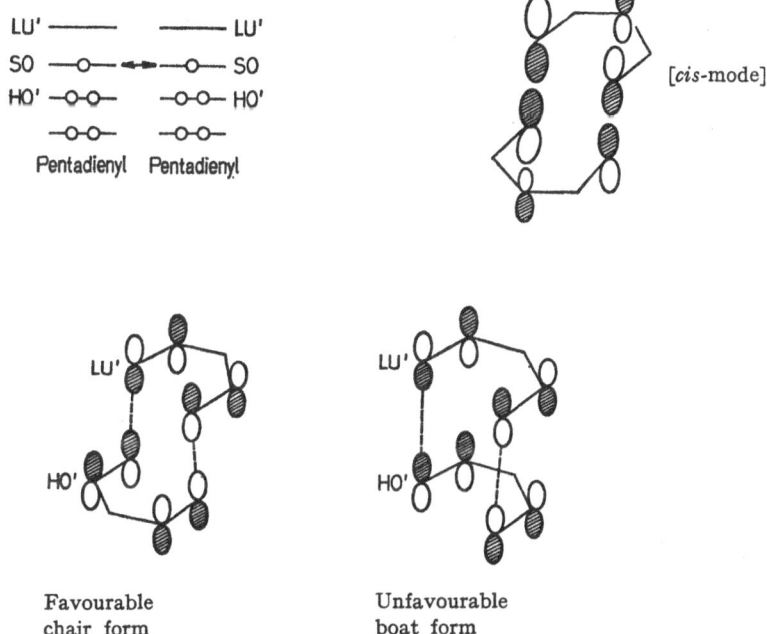

Pentadienyl Pentadienyl

[*cis*-mode]

Favourable   Unfavourable
chair form   boat form

Fig. 7.29b. The transition state in the Cope rearrangements. Extended Cope rearrangement

sign of the hydrogen atom comes to have the same sign as the *upper* half (the *same* side part as the H atom with respect to the molecular plane) of the π AO of the other end of butadiene. The "selection rule" derived from Eq. (7.1) is thus satisfied so that the hydrogen can migrate "suprafacially". The thermal 1,5-migrations are experimentally known. On the contrary, thermal 1,7-transfer does not fulfil this requirement. Except in very special cases [141], such "antarafacial" displacements are not actually known. The relation is reversed in the case of photochemical processes which are considered to occur in the lowest excited state of the polyene π part. The 1,7-migration is favourable in this case. Many experimental evidences are mentioned in reference [51].

a) Ground-State Reaction

i) 1,5-migration
   (in general
   1,(4n + 1)-transfer)

ii) 1,7-migration
   (in general
   1,(4n + 3)-transfer)

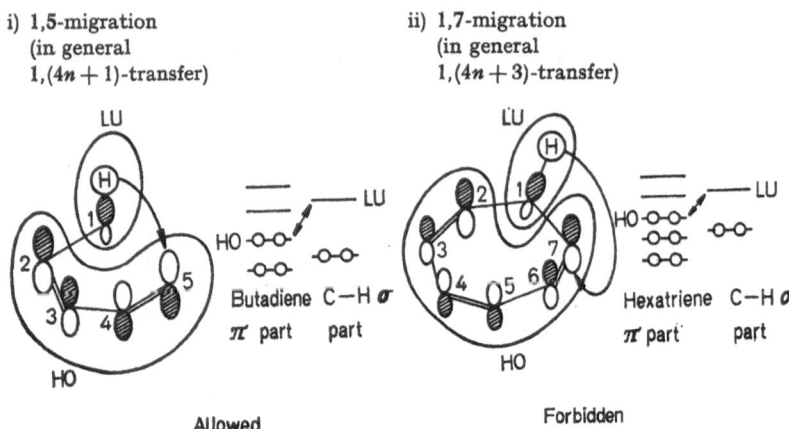

b) Excited-State Reaction

i) 1,7-migration
   (1,4n + 1)-transfer)

ii) 1,5-migration
   (1,(4n + 3)-transfer)

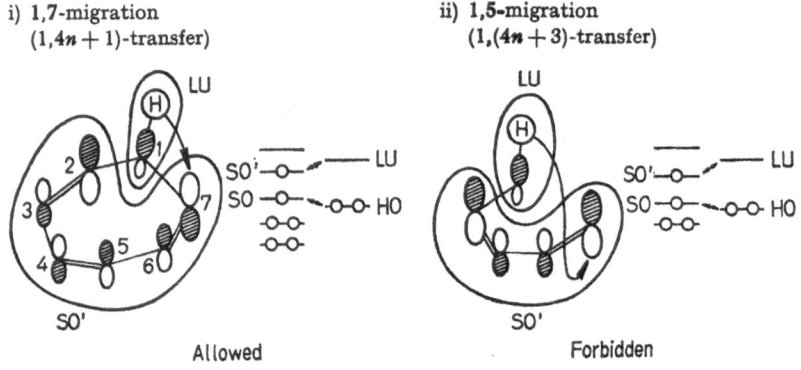

Allowed   Forbidden

(The SO—HO interaction also leads   (The SO—HO interaction also leads
to the same conclusion)   to the same conclusion)

Fig. 7.30. The selection rule for the hydrogen migration in olefins

The cyclization of conjugated polyenes and the inverse reaction were those processes which provided superb materials [142] leading to the Woodward-Hoffmann rule [51].

The energy change, ΔE, due to a *new "bond"* arising between two $2p\pi$ AO's of a conjugated hydrocarbon, r and s, is simply represented by the Hückel calculation as

$$\Delta E = 2\, P_{rs}\, \gamma \qquad (7.2)$$

in which P_{rs} is the π bond order between rth and sth π OA's and γ is the "resonance" integral between these two AO's. Of course, the sign of γ depends on the sign of π AO's adopted. If the signs of π AO's are taken, as used to be, as indicated in Fig. 7.31 a, where each $2\,p\pi$ AO has the same sign in the same side of molecular plane, there may be two possible cases of interaction between two π AO's, which are illustrated in Fig. 7.31 b. In Type I interaction the two π AO's overlap in the region of the same sign, whereas Type II overlapping is concerned with the regions of the different signs. In the former case the resonance integral is negative, while in the latter case it becomes positive. Eq. (7.2) shows that the

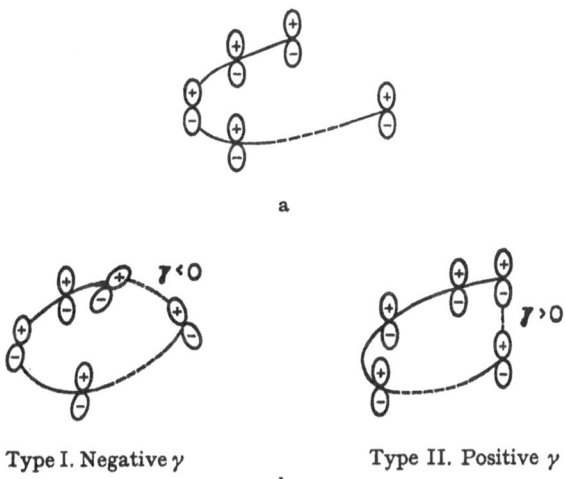

a

Type I. Negative γ Type II. Positive γ

b

Fig. 7.31a and b. The relation between the mode of cyclization and the sign of π AO's. a) The assignment of signs to AO's in conjugated polyenes, b) Two modes of σ-type interaction between two AO's

stabilization due to this overlapping occurs in the case of $P_{rs} > 0$, when $\gamma < 0$, and in the case of $P_{rs} < 0$, when $\gamma > 0$, since the perturbation in the Hamiltonian operator is considered to be negative. Therefore, the selection of the two modes of interaction depends on the sign of π bond order between the two π AO's.

The sign of P_{rs} with respect to various pairs of positions in conjugated polyenes in their ground state was investigated [143]. The result is indicated in Fig. 7.32. The π bond order is always positive in the cases where

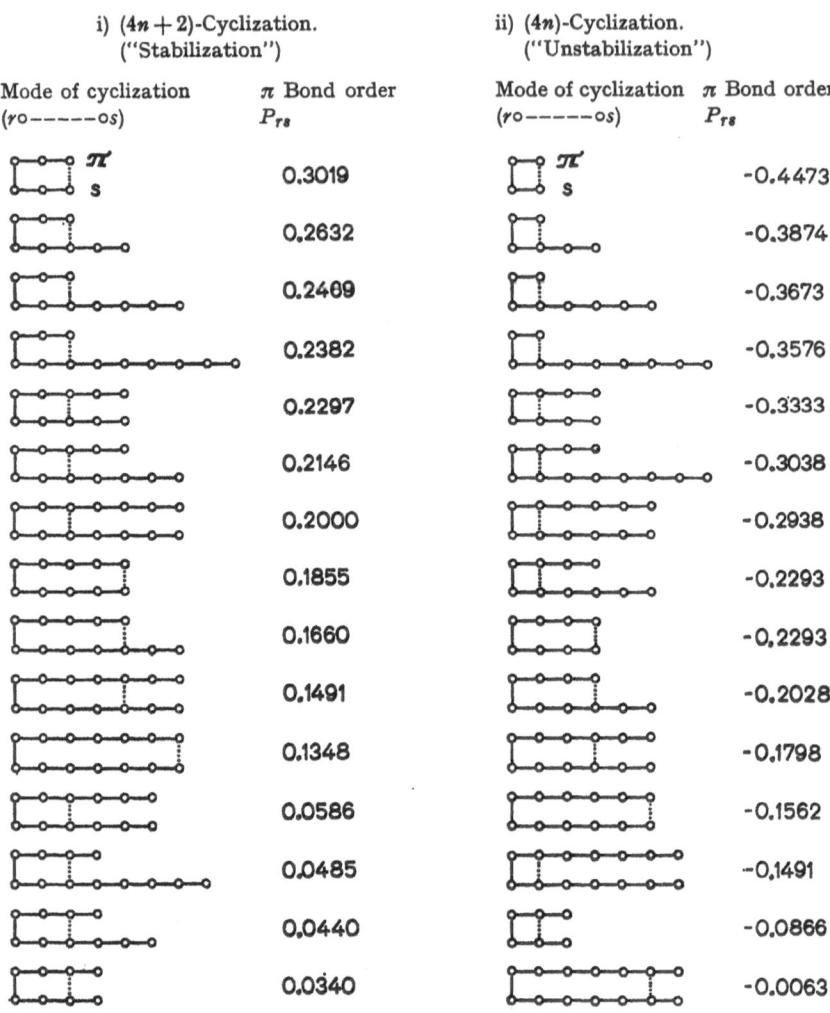

i) (4n + 2)-Cyclization. ("Stabilization")		ii) (4n)-Cyclization. ("Unstabilization")	
Mode of cyclization (ro-----os)	π Bond order P_{rs}	Mode of cyclization (ro-----os)	π Bond order P_{rs}
	0.3019		−0.4473
	0.2632		−0.3874
	0.2469		−0.3673
	0.2382		−0.3576
	0.2297		−0.3333
	0.2146		−0.3038
	0.2000		−0.2938
	0.1855		−0.2293
	0.1660		−0.2293
	0.1491		−0.2028
	0.1348		−0.1798
	0.0586		−0.1562
	0.0485		−0.1491
	0.0440		−0.0866
	0.0340		−0.0063

Fig. 7.32. The cyclization of conjugated polyenes [143]

r—s cyclization might form a $(4n + 2)$-cycle, while it is always negative in the formation of $(4n)$-cycle. Hence, $(4n + 2)$-cyclization would take place by way of Type I interaction, and $(4n)$-cyclization through Type II interaction [62,63]. Mathematical formulations were made for such explanations [62,144].

Woodward and Hoffmann have first disclosed that the thermal $(4n + 2)$-cyclization (and also the photochemical $(4n)$-cyclization) takes place via Type I process, and the photochemical $(4n + 2)$-cyclization (and also the thermal $(4n)$-cyclization) via Type II process [51]. They called the former (Type I) process *"disrotatory"*, while the latter (Type II) process was referred to as *"conrotatory"*. They attributed this difference in selectivity to the symmetry of HO and SO' MO in the ground-state and excited-state polyene molecules, respectively (Fig. 7.33). The former is symmetric with respect to the middle of the chain, and the latter antisymmetric, so that the intramolecular overlapping of the end regions having the *same* sign might lead to the Type I and Type II interactions, respectively.

The reverse process to cyclization, that is, the *ring-opening* of cyclic polyenes was discussed simultaneously by Woodward and Hoffmann [51]. Here we might adopt another way of reasoning which is consistent with the discussions made since the beginning of this section. The mechanism of ring cleavage is understood by considering the participation of the

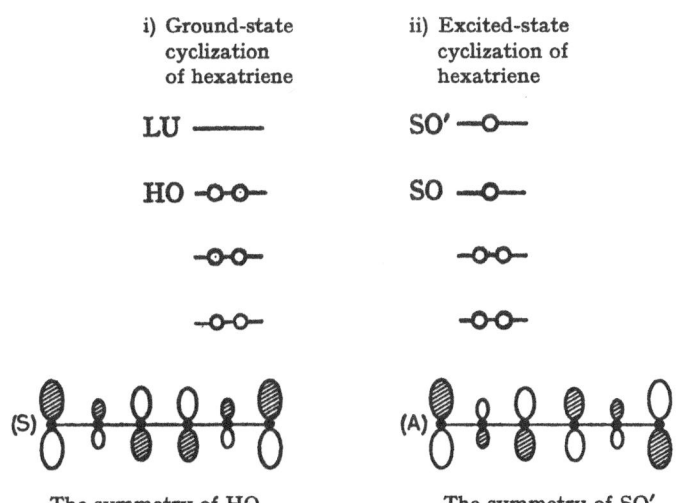

i) Ground-state cyclization of hexatriene

ii) Excited-state cyclization of hexatriene

The symmetry of HO

The symmetry of SO'

Fig. 7.33. The symmetry of the highest occupied MO in the groundstate and the excited-state hexatriene molecules

C—C σ bond to be cleft (Fig. **7.34**). The LU of the C—C σ part will conjugate with the HO of the π part of the ground-state polyene moiety in case of reaction, so that the orbital symmetry relations clearly determines the direction of bond fission. The direction of change is indicated by arrows. In this manner, in the thermal opening, the (**4n**)-chain will be

i) Thermal opening

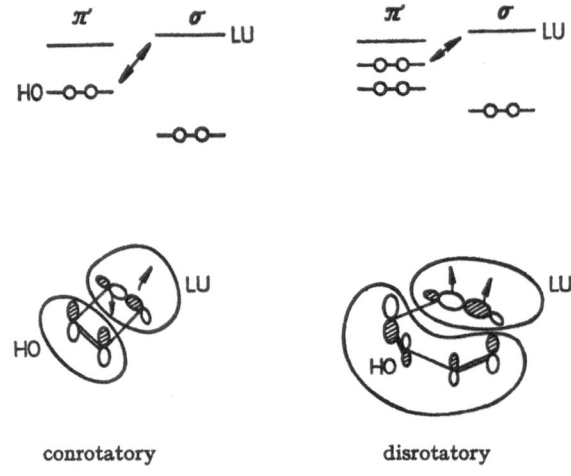

conrotatory   disrotatory

ii) Photochemical opening

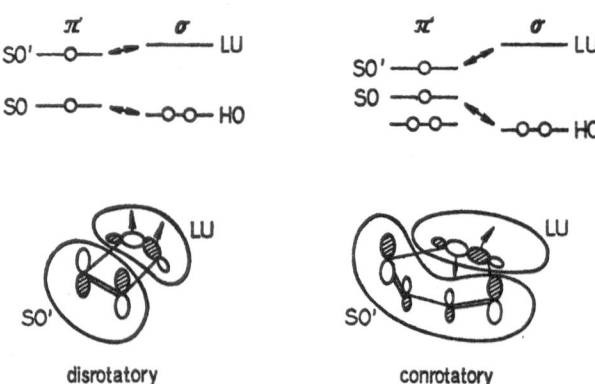

disrotatory   conrotatory

Fig. 7.34. The steric pathway of the ring-opening of cyclic polyenes. (In case ii) the consideration of the SO—HO interaction does not change the conclusion)

formed by a conrotatory pathway, whereas the $(4n+2)$-chain will be produced through a disrotatory process. In the photochemical cleavage, on the contrary, the $(4n)$-chain formation will proceed by a disrotatory fashion and the $(4n+2)$-chain formation by a conrotatory mode. Such a conclusion is most easily derived by the relation of Eq. (7.1), if we investigate the direction of arrows indicated in Fig. 7.34.

It is of interest to investigate the usefulness of this theory to the chemical change involving the interaction between the σ and π parts of conjugated systems [56,62,145]. Such σ-π interactions are frequently stereoselective. The addition to olefinic double bonds and the α,β-elimination are liable to take place with the *trans*-mode [146]. The Diels-Alder reaction occurs with the *cis*-fashion with respect to both diene and dienophile.

The mode of σ-π interaction is classified into *syn*- and *anti*-interactions. These are defined as indicated in Fig. 7.35. The carbon atoms initially sp^3-hybridized change into the sp^x-hybridized state where x is a number between 2 and 3.

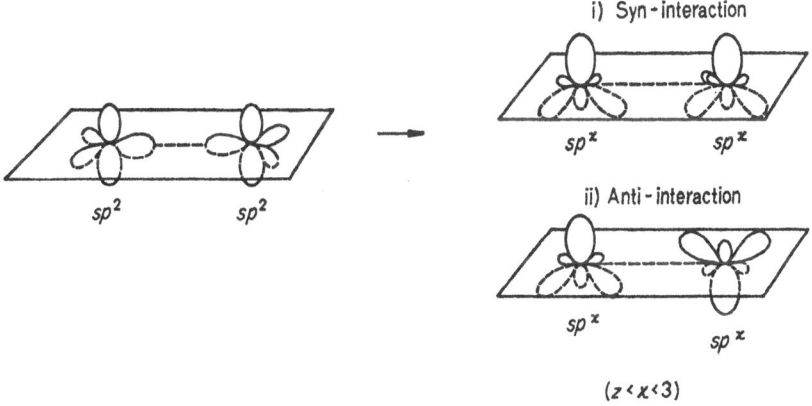

Fig. 7.35. The two modes of a two-centre $\sigma-\pi$ interaction

The *α,β-noncycloaddition to an ethylenic bond* and *α,β-elimination* are taken as the first example. The σ and π parts are regarded as if they were two separate molecules. The direction of change in hybridization is dominated by the overlapping of LU of σ part (Fig. 7.36a) and HO of π part (Fig. 7.36b), so that the mode of interaction becomes as indicated in Fig. 7.36.c.

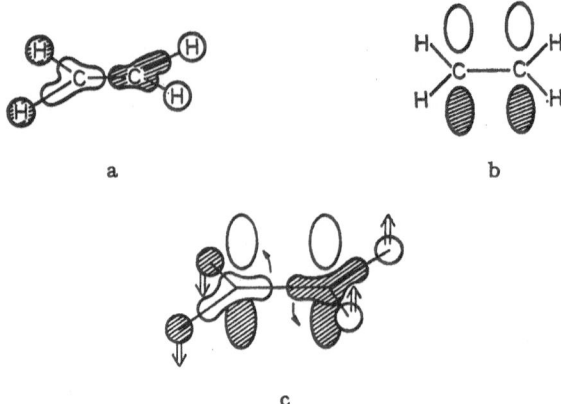

Fig. 7.36a—c. The α,β-noncyclic $\sigma-\pi$ interaction. a) σ LU MO, b) π HO MO,
c) The direction of interaction [Anti-mode]

→ : The direction of σ AO mixing
⇒ : The direction of nuclear configuration change

Similar treatment explains the prevalence of the *syn*-mode [147,148)] in
α,γ-interaction in S_N2' reactions (Fig. 7.37a). The α,δ-interaction (e.g.
S_N2' type reaction) is predicted to occur with *syn*-mode, and α,ε-interaction with *anti*-mode (Fig. 7.37b and c).

Also the σ-π interaction in Diels-Alder additions, which occurs with
syn-fashion with regard to both diene and dienophile, is explained (Fig.
7.38). For the first place, the p-σ type interaction is allowed, by the selection rule already mentioned, between the π-part of butadiene and the
π-part of ethylene. Once this weak p-σ type interaction starts, the p AO
part forms a six-electron system. The HO of this p-part will come from
HO of butadiene π-part interacting with LU of ethylene π-part will
interact with σ-LU's of both butadiene and ethylene. The mode of interaction is as indicated in Fig. 7.38.

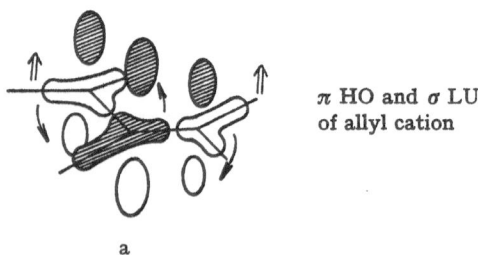

π HO and σ LU
of allyl cation

a

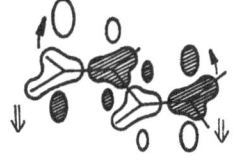

π HO and σ LU
of butadiene

b

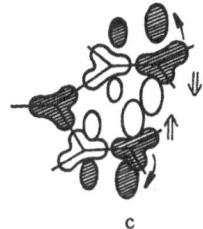

π HO and σ LU
of pentadienyl cation

c

Fig. 7.37a—c. The α,γ-, $\alpha\delta$-, and $\alpha,\varepsilon - \sigma,\pi$-interactions

a) α,γ-Interaction [*Syn*-model]
b) α,δ-Interaction [*Syn*-model]
c) α,ε-Interaction [*Anti*-model]

→ : The direction of σ AO mixing
⇒ : The direction of nuclear configuration change
The hydrogen AO's are not indicated here

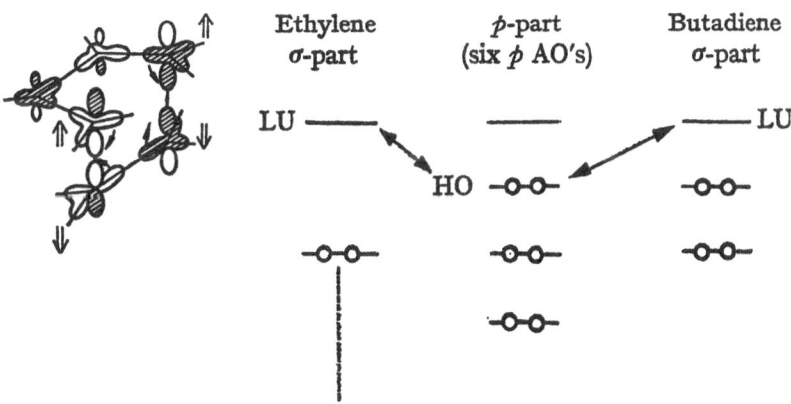

Fig. 7.38. The mode of $\sigma-\pi$ interaction in Diels-Alder reactions

→ : The direction of σ—AO mixing
⇒ : The direction of nuclear configuration change
The hydrogen AO's are not indicated here

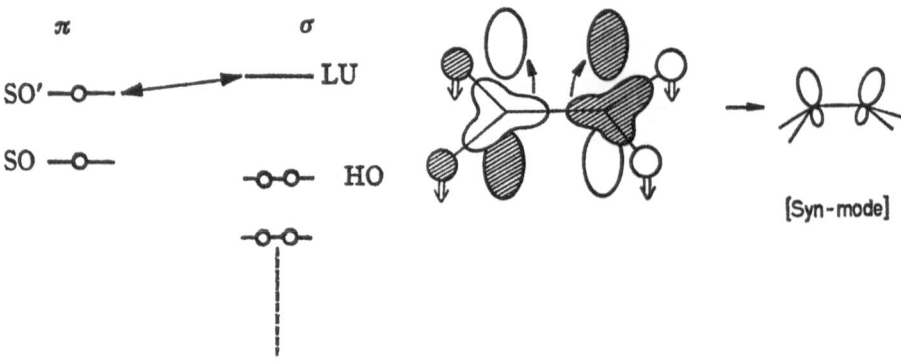

Fig. 7.39. The 1,2-addition to the excited-state ethylenic bond

The σ-π interaction in the excited-state π electron systems is also successfully treated. The 1,2-addition will take place with *cis* mode as is indicated in Fig. 7.39. This was predicated in reference [56]. Experimental evidence [64,149] is the photoinduced addition of *N*-chlorourethane to olefins which gives mainly *cis* addition product, while thermal addition produces a dominantly *trans* adduct.

An interesting example of the application of the theory is a prediction of a new route to *polyamantane* by polymerization of *p*-quinodimethane [121]. The first step would be π-π overlapping interaction. The HO and LU of quinodimethane are indicated in Fig. 7.40a. The mode of π HO-LU interaction and the possible structure of polyamantane derived therefrom (Type I polymer) can be seen in Fig. 7.40b. On the other hand, the direction of the hybridization change would be controlled by the σ-π interaction. The nodal property of π HO and σ LU of the monomeric unit are as shown in Fig. 7.40c, so that the hybridized states of carbon atoms might change into the form illustrated in Fig. 7.40d to lead to the Type II polymer.

Miscellaneous examples of σ-π interactions are listed in the following and in Fig. 7.41. The theoretical conclusion serves in some cases as the explanation of experience in relation to the direction of stereoselection and in some cases as prediction.

a) The HO and LU in the pi-electronic part of *p*-quinodimethane

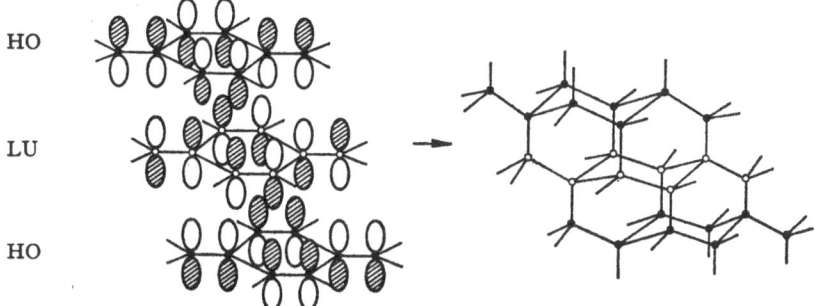

[Type I polymer]

b) The mode of π HO—LU interaction and the possible structure of polyamantane

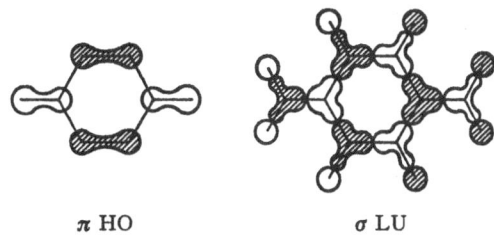

π HO   σ LU

c) The nodal property of π HO and σ LU

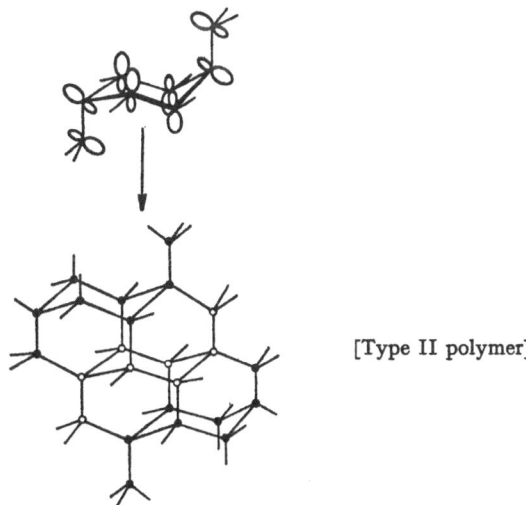

[Type II polymer]

d) The mode of σ—π interaction and the possible structure of polymer

Fig. 7.40a—d. A prediction of a new route to polyamantane on the basis of orbital symmetry consideration. (The shaded and unshaded areas correspond to the positive and negative regions of the wave functions, respectively)

a) Cope and Claisen rearrangements

b) Bis-methylene-cyclobutane rearrangement [150,151)

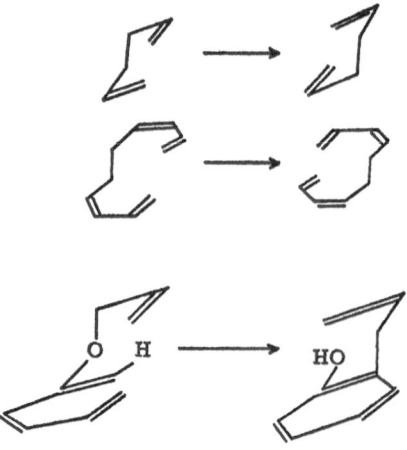

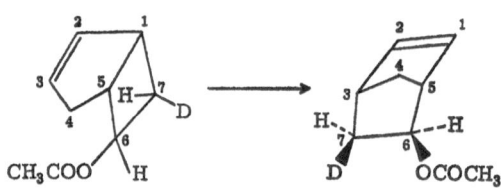

c) Bicycloheptene rearrangement [152)

d) Deamination of cyclic imines [153)

These reactions seem to take place through participation of the σ bond to be split in conjugation with the π part. The direction of the bond fission is indicated in Fig. 7.41 by arrows.

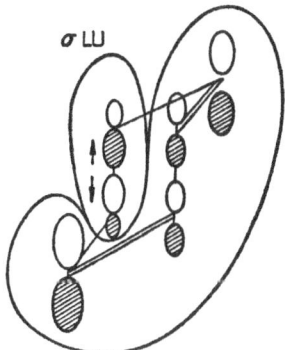

HO of a system consisting of two ethylenes connected by a weak $p\sigma$ bond

a) (i) Cope rearrangement

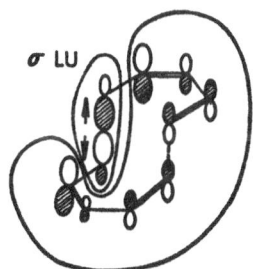

HO of a system consisting of two butadienes connected by a weak $p\sigma$ bond

a) (ii) Extended Cope rearrangement

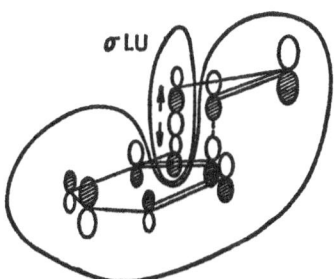

HO of a system consisting of ethylene and benzene connected by a weak $p\sigma$ bond

a) (iii) Claisen rearrangement

77

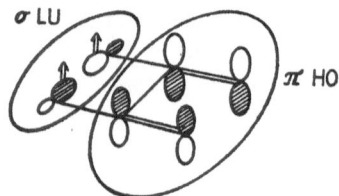

(i) Ground-state reaction

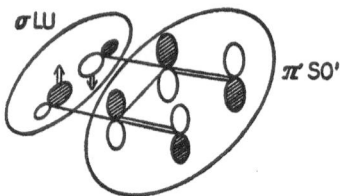

(ii) Excited-state reaction

b) Bis-methylene-cyclobutane rearrangement

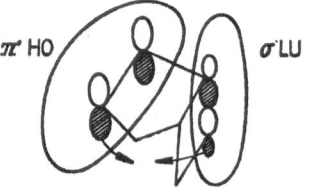

c) Bicycloheptene rearrangement

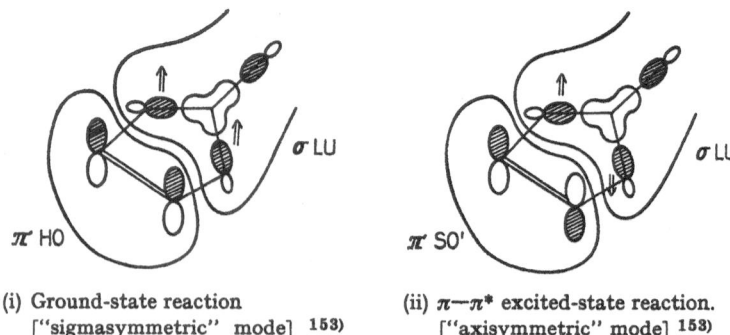

(i) Ground-state reaction
["sigmasymmetric" mode] [153)

(ii) $\pi-\pi^*$ excited-state reaction.
["axisymmetric" mode] [153)

d) Deamination of cyclic imines

Fig. 7.41a—d. Various examples of $\sigma-\pi$ interaction. (Arrows indicate the direction of σ bond fission)

8. The Nature of Chemical Interaction

As has been thoroughly discussed in Chap. 3, the chemical interaction between molecular systems is divided into the Coulomb interaction, the exchange interaction, the charge-transfer interaction, and the polarization interaction. By way of the charge-transfer interaction and the polarization interaction, the different electron configurations come to be mixed with the initial one. This is a chemical excitation process. The molecular shape will tend to change so as to take on a more stable nuclear configuration. That is to say, the change in the electronic state impels the nuclei to rearrange themselves.

Among the two sorts of interaction mentioned above, one of the reactants happens to be an electron-donor and the other an electron-acceptor, as is the case in most heterolytic reactions. In such cases the charge-transfer effect will perhaps predominate over the polarization effect. Even in homolytic interactions, the importance of the mutual charge transfer is not to be disregarded.

In the majority of cases of *charge-transfer interaction* in which reactants are free to change their nuclear configuration, the HO MO of the donor molecule and the LU MO of the acceptor molecule become most important, if the nuclear configuration change along the reaction pathway is taken into consideration, as has been made clear in Chap. 4.

As the reaction proceeds, each reactant molecule changes the nuclear configuration in the direction of stabilization which conforms to the stable configuration of the donor molecule with one electron taken off HO, or that of the acceptor molecule with one electron added to LU. Such nuclear rearrangement, on the one hand, is accompanied by an unstabilization, like promotion in molecule formation from atoms, which is the principal origin of "activation energy". On the other hand, this nuclear rearrangement further accelerates the HO—LU interaction, each co-operating with the other to facilitate the interaction.

This is the probable mechanism of determination of the reaction pathway. The HO—LU interaction thus helps to promote smooth reaction. In the reactions in which a radical or an excited molecule takes part, the overlapping of the SO MO of the specie with the other reactant is of significance and plays the role of HO or LU of common molecules.

The ease with which the reaction proceeds is directly related to the property or behaviour of these particular MO's connecting these to the phenomena of orientation or stereoselection. The electron distribution (valence-inactive population) plays a leading role in the interaction between the particular orbitals, HO, LU, and SO, in usual molecules, no matter whether they are saturated or unsaturated, and determines the orientation in the molecule in the case of chemical interaction. In that case, the extension and the nodal property of these particular MO's decide the spatial direction of occurrence of interaction.

It is understood that the direct *"motive force"* which drives a sizable molecule, even a complicated organic molecule, to chemical reaction may be ascribed to merely *one electron* (or sometimes more) whose mass is less than a ten or hundred thousandth of that of the molecule. In some cases, the existance of a field of "vacant" orbitals extending for long distances facilitates the initial interaction and gives the reagent a chance to select the reaction path.

The nature of intermolecular force is essentially no different from that which participates in the chemical bond or chemical reaction. The factor which determines the stable shape of a molecule, the influence on the reaction of an atom or group which does not take any direct part in the reaction, and various other sterically controlling factors might also be comprehended by a consideration based on the same theoretical foundation.

The secondary or tertiary structure of high polymers, the catalytic effect of organic molecules in majority of enzymatic reactions, and common chemical interactions in heterogeneous systems may also be controlled by factors of the same category. The high selectivity observed in those sorts of interaction might originate from the selectivity of the molecular field which is formed by the complicated molecular systems involved. The possibility afforded by such a molecular field to possess any cause of selectivity in the chemical interaction will easily be recognized by the previous discussions.

9. References

1) Born, M., Oppenheimer, J. R.: Ann. Physik *84*, 457 (1927).
2) Hartree, D. R.: Proc. Cambridge Phil. Soc. *24*, 89, 111 (1928).
3) Fock, V.: Z. Physik *61*, 126 (1930).
4) Kolos, W., Roothaan, C. C. J.: Rev. Mod. Phys. *32*, 219 (1960).
5) — Wolniewicz, L.: Rev. Mod. Phys. *35*, 473 (1963).
6) Buenker, R. J., Peyerimhoff, S. D., Whitten, J. L.: J. Chem. Phys. *46*, 2029 (1967).
7) Wahl, A. C., Bertoncini, P. J., Das, G., Gilbert, T. L.: Intern. J. Quant. Chem. *1*, 103 (1967).
8) Ritchie, C. D., King, H. F.: J. Chem. Phys. *47*, 564 (1967).
9) Pitzer, R. M.: J. Chem. Phys. *47*, 965 (1967).
10) Roothaan, C. C. J.: Rev. Mod. Phys. *23*, 69 (1951).
11) Slater, J. C.: Phys. Rev. *36*, 57 (1930).
12) Boys, S. F.: Proc. Roy. Soc. (London) *A 200*, 542 (1950).
13a) Higuchi, J.: J. Chem. Phys. *22*, 1339 (1954).
13b) Karo, A. M.: J. Chem. Phys. *30*, 1241 (1959).
14) Löwdin, P.-O.: J. Chem. Phys. *18*, 365 (1950).
15) Clementi, E.: J. Chem. Phys. *36*, 33 (1962).
16) Hollister, C., Sinanoğlu, O.: J. Am. Chem. Soc. *88*, 13 (1966).
17) Snyder, L. C., Basch, H.: J. Am. Chem. Soc. *91*, 2189 (1969).
18) Roothaan, C. C. J.: Rev. Mod. Phys. *32*, 179 (1960).
19) Amos, A. T., Hall, G. G.: Proc. Roy. Soc. (London) *A 263*, 483 (1961).
20) Das, G., Wahl, A. C.: J. Chem. Phys. *44*, 87 (1966). — Das, G.: J. Chem. Phys. *46*, 1568 (1967).
21) Pople, J. A.: Trans. Faraday Soc. *49*, 1375 (1953).
22) Kon, H.: Bull. Chem. Soc. Japan *28*, 275 (1955).
23) Pariser, R., Parr, R. G.: J. Chem. Phys. *21*, 466, 767 (1953).
24) Yonezawa, T., Yamaguchi, K., Kato, H.: Bull. Chem. Soc. Japan *40*, 536 (1967).
25) Kato, H., Konishi, H., Yamabe, H., Yonezawa, T.: Bull. Chem. Soc. Japan *40*, 2761 (1967).
26) Yonezawa, T., Nakatsuji, H., Kato, H.: J. Am. Chem. Soc. *90*, 1239 (1968).
27) — Kato, H., Kato, H.: Theoret. Chim. Acta *13*, 125 (1969). — Kato, H , Kato, H., Konishi, H., Yonezawa, T.: Bull. Chem. Soc. Japan *42*, 923 (1969).
28) Pople, J. A., Santry, D. P., Segal, G. A.: J. Chem. Phys. *43*, S129 (1965). — Pople, J. A., Segal, G. A.: ibid. *43*, S136 (1965); *44*, 3289 (1966).
29) Katagiri, S., Sandorfy, C.: Theoret. Chim. Acta *4*, 203 (1966).
30) Imamura, A., Kodama, M., Tagashira, Y., Nagata, C.: J. Theoret. Biol. *10*, 356 (1966).
31) Hoffmann, R.: J. Chem. Phys. *39*, 1397 (1963).
32) Sandorfy, C.: Can. J. Chem. *33*, 1337 (1955).

References

33) Fukui, K., Kato, H., Yonezawa, T.: Bull. Chem. Soc. Japan *33*, 1197, 1201 (1960).
34) Hückel, E.: Z. Physik *60*, 423 (1930).
35) Pauling, L., Wheland, G. W.: J. Chem. Phys. *1*, 362 (1933).
36) Brown, R. D.: Quart. Rev. *16*, 63 (1952).
37) Ri, T., Eyring, H.: J. Chem. Phys. *8*, 433 (1940).
38) Pullman, A., Pullman, B.: Experientia 2, 364 (1946).
39) Dewar, M. J. S.: Trans. Faraday Soc. *42*, 764 (1946).
40a) Coulson, C. A.: Discussions Faraday Soc. *2*, 9 (1947); J. Chim. Phys. *45*, 243 (1948).
40b) — Longuet-Higgins, H. C.: Proc. Roy. Soc. (London) *A 191*, 39; *A 192*, 16 (1947).
41) Wheland, G. W.: J. Am. Chem. Soc. *64*, 900 (1942).
42) Dewar, M. J. S.: J. Am. Chem. Soc. *74*, 3357 (1952).
43a) Fukui, K., Yonezawa, T., Shingu, H.: J. Chem. Phys. *20*, 722 (1952).
43b) — — Nagata, C., Shingu, H.: J. Chem. Phys. *22*, 1433 (1954).
44) — Molecular Orbitals in Chemistry, Physics, and Biology, p. 513 (ed. by P.-O. Löwdin and B. Pullman). New York: Academic Press 1964.
45) Nagakura, S., Tanaka, J.: J. Chem. Soc. Japan, Pure Chem. Sect. 75, 993 (1954).
46) Brown, R. D.: J. Chem. Soc. *1959*, 2232.
47) Mulliken, R. S.: J. Am. Chem. Soc. *74*, 811 (1952).
48) — Rec. Trav. Chim. *75*, 845 (1956).
49) Fukui, K., Yonezawa, T., Nagata, C.: Bull. Chem. Soc. Japan 27, 423 (1954); J. Chem. Phys. *27*, 1247 (1957).
50) — Modern Quantum Chemistry. Istanbul Lectures, Part 1, p. 49; O. Sinanoğlu, ed. New York: Academic Press 1965.
51) Woodward, R. B., Hoffmann, R.: J. Am. Chem. Soc. *87*, 395, 2511 (1965).
52) Fukui, K., Fujimoto, H.: Bull. Chem. Soc. Japan *41*, 1989 (1968).
53) Weinbaum, S.: J. Chem. Phys. *1*, 593 (1933).
54) Mulliken, R. S.: J. Chem. Phys. *36*, 3428 (1962).
55) Brillouin, L.: Actualites Sci. Ind. *71* (1933); *159* (1934).
56) Fukui, K., Fujimoto, H.: Bull. Chem. Soc. Japan *39*, 2116 (1966).
57) Mulliken, R. S.: J. Chim. Phys. *46*, 497 (1949).
58) Koopmans, T.: Physica *1*, 104 (1933).
59) Fukui, K., Fujimoto, H.: Bull. Chem. Soc. Japan *42*, 3399 (1969).
60) Rudenberg, K.: Rev. Mod. Phys. *34*, 326 (1962).
61) Hine, J.: J. Org. Chem. *31*, 1236 (1966) and references cited therein.
62) Fukui, K.: Bull. Chem. Soc. Japan *39*, 498 (1966).
63) — Tetrahedron Letters *1965*, 2009.
64) — Fujimoto, H.: Mechanisms of Molecular Migrations, Vol. 2 (B. S. Thyagarajan, ed.). Wiley-Interscience 1969.
65) Streitwieser, Jr., A.: Molecular Orbital Theory for Organic Chemists. New York: John Wiley & Sons, Inc. 1961.
66) Dewar, M. J. S.: The Molecular Orbital Theory of Organic Chemistry. New York: McGraw-Hill Book Co. 1969.
67) Fukui, K.: J. Chem. Soc. Japan (Ind. Chem. Sect.) *69*, 794 (1966).
68) — Kato, H., Yonezawa, T.: Bull. Chem. Soc. Japan *34*, 1112 (1961).
69) — Yonezawa, T., Nagata, C.: J. Chem. Phys. *27*, 1247 (1957).
70) Mulliken, R. S.: J. Chem. Phys. *23*, 1833, 1841, 2338, 2343 (1955).

[71] Ruedenberg, K.: Rev. Mod. Phys. *34*, 326 (1962).

[72] Fukui, K., Fujimoto, H.: Tetrahedron Letters *1965*, 4303.

[73] Brown, R. D.: J. Chem. Soc. *1959*, 2232.

[74] Hückel, E.: Z. Physik *76*, 628 (1932).

[75] Moffitt, W.: Proc. Roy. Soc. (London) *A 200*, 414 (1950).

[76] Walsh, A. D.: J. Chem. Soc. *1953*, 2260, 2266, 2288.

[77] Fukui, K., Imamura, A., Yonezawa, T., Nagata, C.: Bull. Chem. Soc. Japan *34*, 1076 (1961).

[78] Huo, W. M.: J. Chem. Phys. *43*, 624 (1965).

[79] Brion, H., Moser, C.: J. Chem. Phys. *32*, 1194 (1960).

[80] Hazelrigg, Jr., M. J., Politzer, P.: J. Phys. Chem. *73*, 1008 (1969).

[81] Shriver, D. F.: J. Am. Chem. Soc. *85*, 1405 (1963).

[82] Bulgakov, N. N., Borisow, Y. A.: Kinetika i Kataliz *7*, 608 (1966).

[83] Fukui, K., Morokuma, K.: Proc. Intern. Symposium Mol. Structure and Spectroscopy, Maruzen & Co. 1962, B — 121 — 1.

[84] Clementi, E.: J. Chem. Phys. *46*, 4731 (1967).

[85a] Al-Jaboury, M. T., Turner, D. W.: J. Chem. Soc. *1964*, 4438.

[85b] Turner, D. W.: Tetrahedron Letters *1967*, 3419.

[86] Bene, J. D., Jaffe, H. H.: J. Chem. Phys. *48*, 1807 (1968).

[87] Kato, Hi., Kato, Ha., Konishi, H., Yonezawa, T.: Bull. Chem. Soc. Japan *42*, 923 (1969).

[88] Tsubomura, H.: Bull. Chem. Soc. Japan *26*, 304 (1953).

[89] Buenker, R. J., Peyerimhoff, S. D., Whitten, J. L.: J. Chem. Phys. *46*, 2029 (1967).

[90] Pitzer, R. M.: J. Chem. Phys. *47*, 965 (1967).

[91] Petke, J. D., Whitten, J. L.: J. Am. Chem. Soc. *90*, 3338 (1968).

[91]' Lukina, M. Y.: Russ. Chem. Rev. *1962*, 419.

[92] Pittman, Jr., C. U., Olah, G. A.: J. Am. Chem. Soc. *87*, 2998, 5123 (1965).

[93] Closs, G. L., Klinger, H. B.: J. Am. Chem. Soc. *87*, 3265 (1965).

[94] Deno, N. C., Richey, H. G., Liu, J. S., Lincoln, D. N., Turner, J. O.: J. Am. Chem. Soc. *87*, 4533 (1965).

[95] Walsh, A. D.: Trans. Faraday Soc. *45*, 179 (1949).

[96] Hoffmann, R.: Tetrahedron Letters *1965*, 3819.

[97] Palke, W. E., Lipscomb, W. N.: J. Am. Chem. Soc. *88*, 2384 (1966).

[98a] Kato, H., Yamaguchi, K., Yonezawa, T., Fukui, K.: Bull. Chem. Soc. Japan *38*, 2144 (1965).

[98b] — Yamaguchi, K., Yonezawa, T.: Bull. Chem. Soc. Japan *39*, 1377 (1966).

[99] Wilke, G.: Angew. Chem. *73*, 581 (1960).

[100] Chatt, J., Venanzi, L. M.: J. Chem. Soc. *1957*, 4735.

[101] Calderon, N., Chen, H. Y., Scott, K. W.: Tetrahedron Letters *1967*, 3327.

[102] Ohorodnyl, H. O., Santry, D. P.: J. Am. Chem. Soc. *91*, 4711 (1969).

[103] Kato, H., Yonezawa, T., Morokuma, K., Fukui, K.: Bull. Chem. Soc. Japan *37*, 1710 (1964).

[104] Chan, A. C. H., Davidson, E. R.: J. Chem. Phys. *49*, 727 (1968).

[105] Fujimoto, H., Fukui, K.: Tetrahedron Letters *1966*, 5551.

[106] Kooyman, E. C., Vegter, G. C.: Tetrahedron *4*, 382 (1958).

[107] Fujimoto, H., Fukui, K.: unpublished paper.

[108] Fukui, K., Morokuma, K., Yonezawa, T.: Bull. Chem. Soc. Japan *34*, 1178 (1961).

[109] Klopman, G., Hudson, R. F.: Theoret. Chim. Acta *8*, 165 (1967).

[110] Hudson, R. F., Klopman, G.: Tetrahedron Letters *1967*, 1103.

References

[111] Klopman, G.: J. Am. Chem. Soc. *90*, 223 (1968).
[112] Salem, L.: J. Am. Chem. Soc. *90*, 543 (1968).
[113] — J. Am. Chem. Soc. *90*, 553 (1968); Chem. Brit. *1969*, 449.
[114] Clementi, E., Clementi, H., Davis, D. R.: J. Chem. Phys. *46*, 4725 (1967).
[115] Paudler, W. W., Blewitt, H. L.: J. Org. Chem. *30*, 4081, 4085 (1965); *31*, 1295 (1966).
[116] Streitwieser, Jr., A., Fahey, R. C.: J. Org. Chem. *27*, 2352 (1962).
[117] Fukui, K.: Report Intern. Symp. Atom. Molec. Quant. Theory, Sanibel Island, Florida, Jan. 1964, p. 61.
[118] Dewar, M. J. S.: Advan. Chem. Phys. *8*, 65 (1965).
[119] — The Molecular Orbital Theory of Organic Chemistry, p. 331. New York: McGraw-Hill 1969.
[120] Fujimoto, H: unpublished paper.
[121] — Kitagawa, Y., Hao, H., Fukui, K.: Bull. Chem. Soc. Japan *43*, 52 (1970).
[122] Kato, H., Morokuma, K., Yonezawa, T., Fukui, K.: Bull. Chem. Soc. Japan *38*, 1749 (1965).
[123] Fukui, K., Fujimoto, H.: Tetrahedron Letters *1965*, 4303.
[124] Fujimoto, H.: unpublished paper.
[125] Fukui, K., Hao, H., Fujimoto, H.: Bull. Chem. Soc. Japan *42*, 348 (1969).
[126] Kwart, H., Takeshita, T., Nyce, J. L.: J. Am. Chem. Soc. *86*, 2606 (1964).
[127] Goering, H. L., Nevitt, T. D., Silversmith, E. F.: J. Am. Chem. Soc. *77*, 4042 (1955).
[128] Stork, G., White, W. N.: J. Am. Chem. Soc. *78*, 4609 (1956).
[129] Fukui, K.: Nippon Kagaku Sen-i Kenkyusho Koenshu *23*, 75 (1966) (in Japanese).
[130] Trevoy, L. W., Brown, W. C.: J. Am. Chem. Soc. *71*, 1675 (1949).
[131] Hao, H., Fujimoto, H., Fukui, K.: Bull. Chem. Soc. Japan *42*, 1256 (1969).
[132] Fujimoto, H., Oba, H., Fukui, K.: Nippon Kagaku Zasshi *90*, 1005 (1969) (in Japanese).
[133] Kita, S., Fukui, K.: Nippon Kagaku Zasshi *88*, 996 (1967) (in Japanese).
[134] Sagawa, S., Furukawa, J., Yamashita, S.: J. Chem. Soc. Japan (Ind. Chem. Sect.) *71*, 1897, 1900, 1909, 1913, 1919 (1968) (in Japanese).
[135] Clar, E.: Polycyclic Hydrocarbons, I and II. Academic Press 1964.
[136] Hoffmann, R., Woodward, R. B.: J. Am. Chem. Soc. *87*, 4388 (1965).
[137] Herndorn, W. C., Hall, L. H.: Tetrahedron Letters *1967*, 3095 and others.
[138] Fukui, K., Fujimoto, H.: Tetrahedron Letters *1966*, 251.
[139] Marvell, E. N., Stephenson, J. L., Ong, J.: J. Am. Chem. Soc. *87*, 1267 (1965) and many other papers.
[140] Doering, W. von E., Roth, W. R.: Tetrahedron *18*, 67 (1962); Angew. Chem. *75*, 27 (1963) and many other papers.
[141] Schlatmann, J. L. M. A., Pot, J., Havinga, E.: Rec. Trav. Chim. *83*, 1173 (1964).
[142] Havinga, E., Schlatmann, J. L. M. A.: Tetrahedron *15*, 146 (1961).
[143] Fukui, K., Imamura, A., Yonezawa, T., Nagata, C.: Bull. Chem. Soc. Japan *33*, 1591 (1960).
[144] — Fujimoto, H.: Bull. Chem. Soc. Japan *40*, 2018 (1967).
[145] — Tetrahedron Letters *1965*, 2427.
[146] Bohm, B. A., Abell, P. I.: Chem. Rev. *62*, 599 (1962).
[147] Goering, H. L., Nevitt, T. D., Silversmith, E. F.: J. Am. Chem. Soc. *77*, 4042 (1955).
[148] Stork, G., White, W. N.: J. Am. Chem. Soc. *78*, 4609 (1956).

[149] Schrage, K.: Tetrahedron Letters *1966*, 5795.
[150] Gajewski, J. J., Shih, C. N.: J. Am. Chem. Soc. *89*, 4532 (1967).
[151] Doering, W. von E., Dolbier, W. R., Jr.: J. Am. Chem. Soc. *89*, 4534 (1967).
[152] Berson, J. A.: Accounts Chem. Res. *1*, 152 (1968).
[153] Lemal, D. M., McGregor, S. D.: J. Am. Chem. Soc. *88*, 1335 (1966).
[154] Fukui, K.: Sigma Molecular Orbital Theory (O. Sinanoğlu and K. B. Wiberg, ed.). The Yale University Press 1969.
[155] Goldstein, M. J.: J. Am. Chem. Soc. *89*, 6357 (1967).
[156] Gloux, J., Guglielmi, M., Lemaire, H.: Mol. Phys. *17*, 425 (1969).

Received December 19, 1969

ISBN 978-3-540-04820-6   ISBN 978-3-540-36200-5 (eBook)
DOI 10.1007/978-3-540-36200-5

Titel-Nr. 7722

SPRINGER-VERLAG
BERLIN·HEIDELBERG·NEW YORK

HMO
Hückel Molecular
Orbitals

Von **E. Heilbronner** und **P. A. Straub**
Laboratorium für Organ. Chemie der
Eidgenössischen Technischen Hochschule
Zürich

816 Seiten. 1966
Loseblattheftung
DM 72,—
US $ 24.00

Die Tabellen enthalten die Eigenwerte, Linearkombinationen, Ladungs- und Bindungsordnungen sowie die Polarisierbarkeit einer Auswahl von π-Elektronensystem-Modellen, berechnet nach dem Hückelschen Näherungsverfahren.
Sie liefern diejenige Information, die für eine Verwendung der HMOs innerhalb einfacher Störungsrechnungen notwendig ist und sollen dem theoretisch interessierten Chemiker einen Grundstock von HMOs zugänglich machen, mit dessen Hilfe ein großer Teil der in der Praxis und während des Studiums auftretenden Probleme gelöst werden kann.
Ein Maschinenprogramm zur selbständigen Berechnung komplizierterer Beispiele ist beigefügt.

SPRINGER-VERLAG
BERLIN·HEIDELBERG·NEW YORK

THEORETICA CHIMICA ACTA

edenda curat: **Hermann Hartmann**

adiuvantibus: C. J. Ballhausen, København;
R. D. Brown, Clayton; E. Heilbronner, Basel;
J. A. A. Ketelaar, Amsterdam; M. Kotani, Tokyo;
J. Koutecký, Praha; J. W. Linnett, Cambridge;
E. E. Nikitin, Moskwa; R. G. Pearson, Evanston;
B. Pullman, Paris; K. Ruedenberg, Ames;
C. Sandorfy, Montreal; M. Simonetta, Milano;
O. Sinanoğlu, New Haven

"Theoretica Chimica Acta" will publish papers dealing with the relationship of chemical and physical phenomena to the deductions made from valence and electronic theories. First consideration will be given to those which are primarily of chemical interest.

1970: 3 volumes (Vol. Nos. 16—18)
Price per volume DM 96,—; US $ 26.40
plus postage

SPRINGER-VERLAG
BERLIN·HEIDELBERG·NEW YORK

STRUCTURE AND BONDING

Editors: P. Hemmerich, Konstanz;
C. K. Jørgensen, Genève; J. B. Neilands,
Berkeley; Sir Ronald S. Nyholm, London;
D. Reinen, Bonn; R. J. P. Williams, Oxford

Vol. 7

With 45 figures
III, 154 pages
1970
Soft cover DM 38,—
US $ 10.50

■ **Prospectus
on request**

The Spectra of Ferric Haems and Haemo-
proteins. By Dr. D. W. Smith, Chemistry
Dept., The University of Sheffield, and Prof.
R. J. P. Williams, Inorganic Chemistry
Laboratory, Oxford

The Absolute Configuration of Transition
Metal Complexes. By Dr. R. D. Gillard and
Dr. P. R. Mitchell, Inorganic Chemistry
Laboratory, The University, Canterbury, Kent

The Application of Nuclear Quadrupole
Resonance Spectroscopy to the Study of
Transition Metal Compounds. By Dr. W. van
Bronswyk, William Ramsey and Ralph Forster
Laboratories, University College, Gower
Street, London W.C. 1

Kationenverteilung zweiwertiger $3d^n$-Ionen
in oxidischen Spinell-, Granat- und anderen
Strukturen. Von Dr. D. Reinen, Anorganisch-
Chemisches Institut der Universität Bonn

NMR

Basic Principles and Progress
Grundlagen und Fortschritte
Editors: **P. Diehl, E. Fluck, R. Kosfeld**

Vol. 1

NMR Studies of Molecules Oriented in the Nematic Phase of Liquid Crystals

By Professor Dr. **P. Diehl** and Dr. **C. L. Khetrapal**

The Use of Symmetry in Nuclear Magnetic Resonance

By Dr. **R. G. Jones**

With 53 fig.
V, 174 pp. 1969
Cloth DM 39,—
US $ 10.80

Part one of this volume contains an introduction to the NMR spectroscopy of molecules dissolved in the nematic phase of liquid crystals. This new type of spectroscopy allows the determination of direct spin-spin coupling constants. In the second part recognition of molecular symmetry and its relation to group theory are shown to be essential components in the understanding of NMR spectra and their analysis.

Vol. 2

NMR-Untersuchungen an Komplexverbindungen

Von Dozent Dr. **H. J. Keller**

Mit 22 Abb.
III, 88 S. 1970
Geb. DM 32,—
US $ 8.80

Der Band gibt einen Überblick über die Anwendungsmöglichkeiten der NMR-Methode auf dia- bzw. paramagnetische Komplexe. Der Verfasser geht auf die große Bedeutung der NMR-Spektroskopie bei der Untersuchung schnell ablaufender Reaktionen ein.

SPRINGER-VERLAG
BERLIN·HEIDELBERG·NEW YORK

In kritischen Übersichten werden in dieser Reihe Stand und Entwicklung aktueller chemischer Forschungsgebiete beschrieben. Sie wendet sich an alle Chemiker in Forschung und Industrie, die am Fortschritt ihrer Wissenschaft teilhaben wollen.

In der Regel werden nur Beiträge veröffentlicht, die ausdrücklich angefordert worden sind. Schriftleitung und Herausgeber sind aber für ergänzende Anregungen und Hinweise jederzeit dankbar. Manuskripte können in den „Fortschritten der chemischen Forschung" in Deutsch oder Englisch veröffentlicht werden.

Jedes Heft der Reihe ist auch einzeln käuflich.

This series presents critical reviews of the present position and future trends in modern chemical research. It is addressed to all research and industrial chemists who wish to keep abreast of advances in their subject.

As a rule, contributions are specially commissioned. The editors and publishers will, however, always be pleased to receive suggestions and supplementary information. Papers are accepted for "Topics in Current Chemistry" in either German or English.

Single issues may be purchased separately.